Environmental Pollution 29

Series Editor
J. Trevors, GUELPH, ON, Canada

The Environmental Pollution book series includes current, comprehensive texts on critical national and global environmental issues useful to scientists in academia, industry and government from diverse disciplines. These include water, air, and soil pollution, organic and inorganic pollution, risk assessment, human and environmental health, environmental biotechnology, global ecology, mathematics and computing as related to environmental pollution, environmental modelling, environmental chemistry and physics, biology, toxicology, conservation and biodiversity, agricultural sciences, pesticides, environmental engineering, bioremediation/ biorestoration, and environmental economics. Environmental problems and solutions are complex and interrelated. Complex problems often require complex solutions. The linkage of many disciplines can result in new approaches to old and new environmental problems as well as pollution prevention. This knowledge will assist in understanding, maintaining and improving the biosphere in which we live.
Proposals for this book series can be sent the Series Editor:
Jack Trevors at jtrevors@uoguelph.ca
or the Publishing Editor
Zachary Romano at Zachary.Romano@springer.com

More information about this series at https://link.springer.com/bookseries/5929

Vladimir N. Bashkin • Olga P. Trubitsina

Geoecological and Geopolitical Risks for the Oil and Gas Industry in the Arctic

Challenges and threats

Springer

Vladimir N. Bashkin
Russian Academy of Sciences
Pushchino, Russia

Olga P. Trubitsina
Northern Arctic Federal University
Arkhangelsk, Russia

ISSN 1566-0745 ISSN 2215-1702 (electronic)
Environmental Pollution
ISBN 978-3-030-95909-8 ISBN 978-3-030-95910-4 (eBook)
https://doi.org/10.1007/978-3-030-95910-4

This Springer imprint is published by the registered company Springer Nature Switzerland AG
The registered company address is: Gewerbestrasse 11, 6330 Cham, Switzerland

Preface

This volume presents the issues of geoecological and geopolitical risks in the Arctic region due to the hydrocarbon development. We focus our attention on the Russian part of this region, nevertheless many assumptions and conclusions are quite applicable to the entire Arctic basin. In the beginning, we reveal the ethical and philosophical foundations of understanding environmental equity and justice in relation to the development of Arctic natural resources. Furthermore, we consider various approaches to the moral aspects of making global management decisions in the field of environmental protection and environmental safety. At the same time, special attention is paid to the analysis of modern geopolitical and geostrategic challenges to the development of the Arctic region. The book identifies the key geopolitical factors affecting the sustainable development of the Arctic and analyzes the similarities and differences in the geostrategic positions of the Arctic and non-Arctic states. The main aspect of this book is geoecology in the Arctic. We reveal the need to consider geopolitical challenges in the process of analyzing geoecological risks for oil and gas development projects in the Arctic region. We propose a model of risk analysis, which is based on critical loads of emitted pollutants. The book also focuses on the issues of environmental rating as an indicator of geoecological risk management of Russian oil and gas companies in the Arctic. In the course of risk management of oil and gas projects in the Arctic region, it is necessary to take into account the geopolitical, climatic, ethical, institutional, geoecological, technological, financial, and other features of the implementation of projects. We hope the consideration of geoecological and geopolitical aspects of the development of Arctic resources on the example of Russian oil and gas companies makes it possible to find a consensus on the scale of the entire Arctic region, taking into account the interests of all other national and international companies involved in this process.

Pushchino, Russia — Vladimir N. Bashkin
Arkhangelsk, Russia — Olga P. Trubitsina

Acknowledgments

The given research was supported by the Ministry of Science and Higher Education of the Russian Federation, topic "Biogeochemical processes of transformation of mineral and organic matter of soils at different stages of evolution of the biosphere and technosphere," No 0191-2021-0004. We thank the lieder of the given topic, Andry O. Alekseev, Doctor of Biological Science, Associated Member of Russian Academy of Science, for valid assistance and support of the research undertaken.

At various stages, our colleagues from Northern (Polar) Federal University, Arkhangelsk, Russia; Gazprom Dobycha Yamburg LLC, Gazprom VNIIGAZ LLC; scientific institutions of the Russian Academy of Sciences; M. V. Lomonosov Moscow State University; and other organizations participated in these works. The authors express their gratitude for fruitful cooperation. We would especially like to note the contribution of Dr. Irina V. Priputina (Institute of Physico-Chemical and Biological Problems of Soil Science of the Russian Academy of Sciences) for the joint writing of Chap. 3, and Prof. A.S. Kazak, DrSc (tech.), and Dr. Ratner, PhD, DA, at VNIIGazeconomics for the joint writing of Sects. 5.2 and 5.3. Most of the model calculations and cartographic support of the work were carried out by A.V. Tankanag, PhD (biol.) (Institute of Cell Biophysics of the Russian Academy of Sciences, Pushchino), and their technical design was prepared by R. A. Galiulina (Institute of Fundamental Problems of Biology of the Russian Academy of Sciences). The authors express their special gratitude to all of them.

Introduction

The Arctic usually refers to the waters and lands located north of the Arctic Circle (parallel 66 ° 33′ 39″). Practically, this is the last large territory on our planet that is relatively untouched by anthropogenic impact. This region makes up a sixth of the world's landmass, it is home to about 4 million people dispersed in the territories of eight circumpolar countries: Russia, Sweden, Finland, Iceland, Norway, Denmark (Greenland), Canada, and the USA. This region is replete with biological resources (fish, marine animals) and energy resources (oil, gas). According to geological surveys, there may be about 20% of the world's reserves of these organic deposits (Gramberg et al., 1983; Gramberg & Pogrebnitsky, 1984; Abramov et al., 1992; Gramberg & Laverov, 2000; Kaminsky, 2009; Ross, 2017).

During the Second World War and the subsequent Cold War, the Arctic was already the subject of geopolitical disputes. Currently, the opposition of the circumpolar states is becoming more acute again due to their desire to establish control, first, over new oil and gas fields, which, in the conditions of a changing climate and the emergence of new technologies, are becoming potentially economically profitable. This, accordingly, turns one of the most stable regions of the world into a hotbed of conflicting geopolitical interests and corresponding geopolitical risks (Huebert et al., 2012; Bashkin, 2018).

In particular, the issue of the delimitation of maritime borders is still the subject of acute diplomatic disputes in almost all eight circumpolar states. However, already in the Ilulissat Declaration of 2008, it was agreed that any foreign policy or regional disputes between the Arctic states should be resolved only by legal means within the framework of multilateral and international agreements (Huebert et al., 2012; Collins, 2017a, b).

In particular, the issue of the delimitation of maritime borders has so far been agreed upon, in particular, the right of a state to control coastal waters beyond 200 nautical miles, if it can prove that "the seabed is an extension of its continental shelf" (CHASE, 2013).

It should be noted here that one of the intractable disputes is the question of the ownership of the Lomonosov Ridge, located near the Geographic North Pole. A number of Arctic countries (primarily Russia, also Denmark and Canada) claim this

Ridge, where, according to geological forecasts, unique oil and gas deposits can be expected (Kaminsky, 2009). Although, after lengthy and extensive marine geodetic studies, the aforementioned countries have submitted justifications for their claims to this territory to the UN, it is difficult to imagine that any decision will be made in the coming years.

Border negotiations are also underway on the Lincoln Sea and Hans Island basin between Canada and Denmark, the Beaufort Sea (Canada and the USA). The reason for the disagreement in all cases is the potential reserves of energy resources.

One can also note the positive results of diplomatic negotiations on the example of the demarcation of the border in the Barents Sea between Russia and Norway in 2010, which allowed the development of oil and gas fields in this area to begin (Harding, 2010).

It should be further pointed out that such negotiations are relevant not only between the circumpolar countries. In recent years, the interests and demands of non-Arctic states, such as China and India, who want to solve their own economic and commercial problems in the region, have been acutely manifested.

All this determines the likelihood of geopolitical risks when involving the resources of the Polar region in economic activity.

At the same time, environmental issues are no less significant, and in some cases even more significant. The problems of environmental safety of the Arctic are of particular importance due to the increased vulnerability of the environment and the intensive development of natural resources in the region. Numerous studies by Russian and foreign scientists show that the levels of pollution of the Arctic territories are low compared to other regions of the globe, but the anthropogenic load on the environment in high latitudes increases due to the further development of economic activity in the Arctic zone, including on the continental shelf.

The Arctic Ocean and its shelf seas are of global importance due to their influence on the World Ocean and atmospheric circulation and the presence of unique biological species in their territory (polar bear, narwhal, walrus, beluga, etc.). Shelf seas are the main zones of ice formation affecting the internal structure of the ocean and the quality of its waters. Being the smallest ocean basin in the world, this ocean plays a crucial role in the movement of oceanic waters, because it is connected to the Atlantic and Pacific Oceans and maintains water exchange with them. Taking into account its role in the formation of the deep waters of the Atlantic Ocean, the cyclonic nature of the surface circulation of the waters of the Eurasian basin, and the impact of resource extraction activities, primarily environmental pollution, environmental transformation processes have a transboundary global character. Priority environmental problems of the Arctic regions are associated with the presence of compact "hot spots" in places of intensive resource extraction activity, dominating among all types of impacts. Here, pollution levels are many times higher than background values, natural ecosystems have been disrupted or destroyed, and natural landscapes have been changed.

The bowels of the Arctic and adjacent (circumpolar) territories represent the largest reserve of mineral and energy resources to meet the growing needs of the world economy. Of particular importance is the development of a long-term strategy for conducting geoecological research in the Arctic in order to understand and establish patterns of its state and development (Kaminsky, 2009).

There are three types of main sources of pollutants entering the Arctic marine waters: (1) exogenous sources – river runoff, aeolian demolition, wave abrasion, glacial, ice, and iceberg separation; (2) endogenous sources – substances coming from the bowels of the Earth, for example, removal of petroleum hydrocarbons from the sedimentary strata, methane flows, and gas hydrate outlets; (3) aquapolitic sources – dumping of waste, transportation of petroleum products and toxic substances by sea, development of offshore fields, and the system of global currents.

Ensuring the environmental safety of the management of Arctic territories and the development of deposits of natural resources is one of the main objectives of the Arctic development strategy. Reliable determination of the parameters of the Arctic environment is the basis of design solutions that minimize natural risks. In order to ensure the sustainable development of the Arctic and near-Arctic territories, it is necessary to develop the main ways to solve the most important environmental problems and create modern systems for integrated monitoring of the state of the environment.

To formulate the main environmental problems of the Arctic, briefly, we will quote the points of the UN Environment Programme:

- Climate change, melting of centuries-old ice
- Pollution of the seas by oil products and industrial and transport waste
- Increased fishing and seafood production
- Reducing the diversity and abundance of animals and plants
- Intensive shipping

The increased interest shown by the international community in the practical development of the Arctic territories in recent years is caused by both economic and geopolitical factors and the need of countries to strengthen national security. One of the main players in Arctic issues is Russia. For the Arctic as a region characterized by an increased degree of ecosystem instability, the issues of rational nature management are of particular relevance. Environmental problems of the northern territories are currently an important subject of discussion both in scientific circles and in the production sector, especially oil and gas. The urgency of Arctic development, while simultaneously reducing the negative impact on the environment, leads to the need to search for new approaches to the development of Arctic territories. It is necessary to develop a concept of rational nature management in the Arctic territories, which will contribute to their sustainable industrial and socio-economic development and help reduce the likelihood of geoecological risks and emergencies. It is necessary to analyze the latest initiatives of interested Arctic states on this issue, consider key environmental problems, and assess the availability of technologies in the field of environmental management to reduce risks of various nature. It is important to take into account the behavior of various stakeholders, both at the state level and various companies operating in the Arctic region.

The prospect of extraction of natural resources as well as the possible opening of the Northwest Passage and the Northern Sea Route due to climate change led to the fact that a number of Arctic countries resumed their military presence in the region. Canada, Russia, and Norway have tried to restore their military capabilities, although such military assets are more aimed at demonstrating sovereignty and exercising oversight in the region than at deploying advanced offensive weapons against other Arctic countries.

Currently, we can still say that the management of the Arctic is based on international law, as well as clear norms and rules. Paradoxically, this is the most important reason for the peaceful settlement of disputes in the region. All major non-Arctic states have also declared their commitment to the practice of interaction between countries adopted in the region. We can hope for opportunities for further improvement of these rules in the future, taking into account the increased interest and participation of non-Arctic countries. This will help to transform the threats of emerging geopolitical and geoecological risks into appropriate opportunities for both countries and industrial companies.

Modern trends in the development of large industrial corporations, which orient business towards a responsible attitude to the use of natural resources, the preservation of a favorable environment, and the creation of a positive "ecological image" dictate the need for prevention and economically justified reduction of geoecological risks. The latter arise during the construction and operation of industrial and infrastructure facilities because of the negative impact of economic activity on the natural environment, as a result of the impact of various pollutants present in man-made emissions, effluents, and waste. It is also important to take into account the reverse negative impact of the changing natural environment on the industry itself. For a long time, scientific research in the field of environmental protection from pollutants has been focused on identifying and analyzing the environmental effects of pollutants on public health, the state of biota, or the quality of natural environments, that is, on ascertaining the fact of environmental violations that have already occurred. Whereas, it is much more urgent to identify, prevent, and minimize negative effects and potentially dangerous environmental situations at the stage of design decisions.

To implement such tasks, a risk analysis methodology is increasingly being used, which includes interrelated procedures for identifying environmental and political hazards, quantifying risks, and managing risks. At the same time, the risk management strategy is focused on minimizing man-made impacts on natural complexes, taking into account their sustainability potential in relation to specific pollutants, as well as on optimizing technical solutions and developing programs for restoring natural complexes (Bashkin, 2006a, b; Samsonov et al., 2007a). Such a strategy is possible on the basis of a system analysis and comprehensive consideration of a set of factors that determine specific geoecological risks and have regional specifics. Due to the complex nature of the interactions of modern production systems with the environment, the practical solution of the problems of assessment and management of geoecological risks is carried out, as a rule, using techniques and methods of system analysis, including the decomposition of a multifactorial complex of interactions into separate components described using relatively simple models, and the subsequent synthesis of the results obtained to search for strategic solutions (Samsonov et al., 2007b).

All these provisions are taken into account to the greatest extent in the methodology of risk analysis, which makes it possible to use it to solve environmental problems in many sectors of Arctic production, including the gas and oil industries. Providing scientific and methodological foundations for the management of geoecological risks associated with the emission of pollutants is a significant and urgent task for the gas industry, where the emission of pollutants (primarily nitrogen and sulfur oxides) is represented in almost all sub-sectors, beginning from production, transport, and storage and ending with gas processing.

Accordingly, the topics of this book are:

1. Environmental equity and justice in relation to the development of natural resources in the Arctic
2. Geopolitical risks for oil and gas industry in the Arctic Zone of the Russian Federation
3. Geoecological risks for oil and gas industry in the Arctic Zone of the Russian Federation (analyses and modeling)
4. Environmental rating as an indicator of geoecological risk management of Russian oil and gas companies in the Arctic
5. Management of geopolitical and geoecological risks as a basis for taking into account national and international interests in the development of natural resources in the Arctic

We have carried out research on the qualitative and quantitative assessment of geoecological and geopolitical risks for the Arctic regions during the extraction and transportation of energy resources over the past decades. The generalized results of these studies, conducted in different years, at different sites and with varying degrees of detail, are reflected in this book. We hope that the topic of analysis of geopolitical and geoecological risks in the development of oil and gas resources in the Arctic region and the results of the work presented in the monograph will be of interest to a wide range of specialists in the ambiguous field of the Arctic.

Institute of Physico-Chemical and Biological Problems of Soil Science of the Russian Academy of Sciences
Pushchino, Russia — Vladimir N. Bashkin

Northern (Polar) Federal University
Arkhangelsk, Russia — Olga P. Trubitsina

Contents

About the Authors

Vladimir N. Bashkin is a principal researcher in the Institute of Physical, Chemical, and Biological Problems of Soil Science at Pushchino Research Center, RAS, and Professor of Biological Sciences. His main research interests are biogeochemistry, geoecology, and risk assessment. Vladimir is author of more than 400 works.

Olga P. Trubitsina holds a PhD (geography) and is an assistant professor at Northern (Arctic) Federal University, Arkhangelsk. Her main research interests are geoecology and risk assessment in the Arctic. Olga is author of more than 80 works.

Irina V. Priputina is a leading researcher in soil science in the Institute of Physical, Chemical, and Biological Problems at Pushchino Research Center, RAS, and he holds a PhD (geography) and is an assistant professor. Her main research interests are biogeochemistry of nitrogen and carbon in terrestrial ecosystems, analysis of the state and dynamics in ecosystems using simulation methods, quantitative assessments of risks, and critical loads of pollutants. Irina is author of more than 90 works.

Chapter 1
Environmental Equity and Justice in Relation to the Development of Natural Resources in the Arctic

Abstract In this chapter, the authors attempt to clarify the concepts of fairness and environmental justice, which are often mixed into one concept in the environmental literature. For this purpose, the philosophical roots and historical views on these concepts are considered. The concept of environmental justice is proposed as a broad general concept that includes all issues of justice in the process of making management decisions in the field of environmental safety in the Arctic, including the fairness of the procedure and fairness in distribution. It is also shown that fairness, similar to environmental justice, has ethical and causative justification. Furthermore, both environmental justice and fairness use ethical and causative arguments. The authors stated that it is possible to find the philosophical and theoretical bases for the concepts of environmental justice and fairness, and enables practical application of the resulting conclusions in the discussions on climate change and environmental issues.

Keywords Environmental equity and justice · Ethical and cause-and-effect philosophies · Environmental justice and fairness · Ethical and causative justification

1.1 Introduction

The many ethical concepts of fairness and environmental justice that can be found in the literature on environmental issues require a unified approach to the theoretical and conceptual determination of their place in the environmental ethics. For instance, the concepts of fairness and environmental justice are often merged, despite having different philosophical bases. This hampers the formulation of a global environmental policy and devoids this issue of conceptual clarity. In this chapter, the authors discuss various approaches to the moral aspects of global governance decisions related to environment protection and ecological security define the conceptual and philosophical notions of environmental justice and fairness in

V. N. Bashkin, O. P. Trubitsina, *Geoecological and Geopolitical Risks for the Oil and Gas Industry in the Arctic*, Environmental Pollution 29,
https://doi.org/10.1007/978-3-030-95910-4_1

terms of philosophy, morality and law. This approach forms the perception of environmental justice as a systemic concept that encompasses all legal aspects of governance decisions related to the environment and ecological security and, therefore, represents fairness. This allows a unification of contradictory and incomplete views of environmental justice that exist in the world, particularly in relation to the development of hydrocarbons in the Arctic and its vast offshore areas for the construction and use of transportation routes. For Russia, the Arctic region is a safety factor for its continued social and economic development.

It is undoubtedly important that the words used for describing or communicating any concept correctly express the intentions. If the intentions and the way that they are expressed are contradictory, the arguments will be perpetuated, misunderstanding will pass from one generation to another, and knowledge will become obscure. It is hard to find a concept in the literature on sustainability that would be more contradictory than the concept of fairness and environmental justice. They are often combined and have become a single notion in the literature (Foster, 1993). In many cases, the notions of fairness and environmental justice are applied inconsistently and incongruently, which makes them difficult to conceptualize.

Other authors as well noted the confusion around the notions of environmental justice and fairness. For instance, Prof. V. Been states that the term "fairness" is excessively generic and uncertain and, therefore, calls for "fairness" and "justice"each time reflect someone's ideas of what is fair, just, impartial, etc. (Been, 1993). The environmental justice advocates' inability to define preventive, compensatory, or corrective goals has also been pointed to. The variety of approaches to defining the ethical concepts in the environmental literature necessitates the introduction of unified framework rules. Their absence threatens the efficiency of both the scientific discourse and the global policies in relation to the environment (Arnold, 1998). The authors particularly focus on the fact that fairness and environmental justice have different philosophical bases and therefore are distinct from each other. Let us review these differences from the points of view of philosophy, law, and morality. The analysis of the philosophical basis for the governance decision-making process is expressed in the ethical and cause-and-effect philosophies. This makes it possible to find the philosophical and theoretical bases for the concepts of environmental justice and fairness, and enables practical application of the resulting conclusions in the discussions on climate change and environmental issues.

1.2 Ethical and Cause-and-Effect Philosophies

Ethical judgments are usually based on the two major paradigms of moral philosophy: deontological (law-based) and causative (goal-based) (Dasgupta, 1990). The deontological reasoning places the law above utility (Kant, 1966). This paradigm is centered on the principles of fairness, basic human rights, duties, responsibility, benevolence, and respect for the inalienable basic rights of other people

(Akaah, 1997). The deontological rationale requires that the person who makes decisions should always be guided by good intentions, without considering possible consequences. This presumes that all people have a set of inherent rights, which must never be infringed upon by anyone, even the state, regardless of the consequences (Nozik, 1974). The principle of fairness requires that these rights are respected, or that a response action is taken if they are violated. Thus, the deontological evaluations focus on the morality of actions per se, rather than on possible consequences and harm (Baron, 1994).

This is sometimes objected to by claiming that the deontological theories contend that "the means justify the ends," and whether actions are "right" or "wrong" does not in the least depend on the ultimate or expected outcome (Carter, 2002). This means that the deontological theories stress the values to be adhered to in the decision-making process. However, upon a closer look, this approach assumes that there is no connection whatsoever between the consequences of actions and the actions themselves. It turns out that an action can be justified if it complies with a certain rule, and this alone determines the morality of the action (VandeVeer & Regan, 1987).

I. Kant reflected the deontological position in many works. He defines the duty as a categorical imperative. This refers primarily to the absolute duty to act in accordance with certain rules or principles: "Act in such a manner that the rule of your will can always become a principle of the general law." Essentially, this repeats the ancient dictum: do unto others, as you would have them do unto you. Do what everyone must do. This frequently referenced quote from Kant reflects the basic point of the theory that an action is moral if it follows the rules of conduct in a given situation.

The cause-and-effect paradigm, on the other hand, recognizes the priority of utility over the law (Rawls, 1972). According to this paradigm, whether an action is "right" or "wrong" is determined by the ultimate results. In this theory, the society simply identifies the desired social goals such as improving the overall welfare situation and requires its members to act accordingly. Further behavior and rules are evaluated exclusively based on their implications for achieving the goal of common utility. If there is a conflict between common good and individual or collective rights, then common good must prevail. The goal orientation is summarized in the work by Prof. Dvorkin: "Fairness is defined by the outcome: political decisions result in unfairness, no matter how fair the procedures for making these decisions are, if they deprive people of resources, freedoms, and opportunities" (Dworkin, 1993). Problems may arise when discussing the goals of social welfare and identifying the principal public goal.

The utilitarian attitude is intrinsic to the cause-and-effect paradigm. One of its advocates is J.S. Mill. His works emphasize the need to maximize a certain utility, understood as a certain level of common public good which he refers to as the "Principle of the Greatest Happiness." The author understands happiness as something that creates or produces the most pleasure and/or the least suffering, and this is what the right and morally justified behavior is in a given situation (Carter, 2002). For instance, according to Mill, behavior is right or moral if it brings good (pleasure) to the maximum number of people. The individualistic version of this concept, expressed in the theory of hedonism, looks for the most pleasure or the least suffering for an individual. J.S. Mill himself believed that his philosophy should have a predominantly collective orientation.

The classical utilitarianism is focused on maximizing the well-being of an individual. On the other hand, most cause-and-effect theories prioritize not only the common welfare but also an equal distribution of wealth (Ikeme, 2003). The principle of accounting for different positions, suggested by J. Rawls, requires improving the level of welfare for the least well-to-do individual (Rawls, 1972). Here we can see the non-utilitarian basis of the cause-and-effect theory. It can be traced in attempts to evaluate the level of economic welfare and development, where health, life expectancy, and education level are viewed as consequential to and reflecting the degree of welfare, without considering the income level or other attributes of well-being or contentment (Sen, 2000). All cause-and-effect theories have a common feature, which is that the efficiency of relevant actions and programs is evaluated in terms of their overall outcome.

Thus, the difference between the ethical and cause-and-effect paradigms of the moral philosophy comes down to two characteristics. The first is evaluating the degree of importance assigned to the rights of an individual or groups that the society as a whole is comprised of and the second is the focus placed on the "outcome" in one case and on the fairness of the procedure for achieving the social goal in the other case. Certainly, any requirement is based on the law, if it reflects someone's individual interests and complies with the procedures in place, or on the goal, if it is called for by the outcome and reflects the interests of the society in general. So, what should the relation be between the two different paradigms of the moral philosophy, on the one hand, and environmental justice and fairness, on the other hand?

1.3 Environmental Justice and Fairness

The conflation of environmentalism and the principle of justice is a relatively new phenomenon. The term "environmental justice"originates from and it is widely used in the scientific community that studies the impacts of environmental risks on ethnic minorities. This primarily refers to the situation in the United States, where the issues of land use at local level turned into a racial hate issue in the 1980s. It looks at fair treatment and meaningful involvement of all population, regardless of race, color, national origin or income, in addressing the issues of development, implementation, and enforcement of environmental laws (Bass, 1998). The keystone of this aspect of environmentalism is the fact that certain groups of population are more vulnerable to environmental pollution than other ones (Buchanan & Mathieu, 1986). At the international level, the practice of harmful waste exports for burial to developing countries is also viewed as an environmental unfairness on a global scale (Lipman, 2011). The Basel Convention ratified bans hazardous waste exports from rich countries, except where these are intended for recycling or reuse

(Martinez-Alier, 2001). Thus, the calls for environmental justice are calls for equality (Been, 1993). They seek compensation for existing or imminent threats in the distribution of environmental benefits or expenses, and decision-making based on the presumption of equality of all people (Bullard, 1999).

Therefore, the concept of environmental justice is applicable to both distribution and procedure-related issues. It is also evident that this concept can be rationalized based on both ethnic and causative considerations. Some authors distinguish between environmentally unfair outcome and environmentally unfair behavior (Baden & Coursey, 2002). An unfair outcome is based on causality, since it is a result of behavior. In this case, it is irrelevant whether the behavior itself is fair or not, if it causes an unfair outcome (Williams, 1999). To the contrary, unfairness of behavior in essence is determined by its morality, rather than consequences, which means that it is of an ethical nature. This also dictates certain procedural codes for environmental justice (Camacho, 1998). For instance, environmental justice advocates typically claim that poor segments of population are exposed to the highest environmental risks due to the fact that they are not sufficiently involved in the decision preparation and making processes because of unfair procedural rules. This is explained by the fact that the authority can only ignore certain groups of population when such groups are not represented or not adequately represented and cannot influence the decision-making processes. Thus, the involvement of concerned groups in the decision-making process is essential for protecting the interests of the society in general. In this sense, behavior will be justified not by its consequences, but by compliance with the procedural rules and existing values, which will be an ethical justification for environmental justice.

Similarly, equality, which is one of the key characteristics of environmental justice, can serve as both an ethical and a causative reference. From the ethics point of view, equality is good in itself, as all people are created equal and should have equal rights (Been & Gupta, 1997). It is based on these grounds that the environmental justice advocates speak about the ethnic minorities and the poorest population groups' particular vulnerability to the harmful impacts on the environment and noxious waste emissions, which violates their equal rights to a healthy environment more than it does for the ethnic majority (Martinez-Alier, 2001). For those who adhere to the cause-and-effect approach, equality is only good if it leads to a positive outcome. If such an outcome is not achieved, the resulting condition will be viewed as unfair. For instance, when a study finds that areas surrounding a harmful factory are mostly populated by poorer residents, and it is not a result of voluntary marginalization, but is due to the market economy that negatively affects the polluted surroundings, makes them cheaper and economically more attractive to poorer social groups, this situation will be viewed as unfair because the conditions are not equal for everyone (Lambert & Boerner, 1997). The minor indigenous ethnic groups of the Arctic, which preserve the invaluable historical experience, including environmental experience, of survival in extreme nature conditions, are also vulnerable in this respect.

The practice of environmental justice includes the preventive, punitive, and correctional components. The preventive component consists in keeping the future in mind when making decisions. This can be found in the international law and national environmental policies. For example, Principle 21 of the Stockholm Declaration 1972, which was updated through Principle 2 of the Rio Declaration 1992, recognizes the countries' right to "develop their own resources in accordance with their own environment and development policies," without causing "damage to the environment of other countries or areas outside of their national jurisdictions" (Kolosov, 1997). This provision looks ahead in the future and aims to prevent a negative impact on the environment outside of the national borders, which would be environmentally unfair.

The punitive role of environmental justice consists in discussing at the international level the use of sanctions and other punitive measures for countries that violate the international environmental agreements (Eglin, 2001). However, the issue of effective and conscientious compliance (or non-compliance) should be dealt with before punitive measures are applied at the international level for committing environmental unfairness. Fines are widely used in regulating the environmental relations at the national level. In the United States, the trend is to step up the sanctions, which may even include a prison sentence (Ausubel & Victor, 1998). This may become an effective regulatory tool if there is no political basis for governance decision-making.

Important provisions on the distributive and procedural dimensions of environmental fairness can be rationalized by using ethical and causative arguments and separating the preventive, correctional, and punitive functions. However, fairness depends on the functioning of courts and deals with a complicated distributive choice. The standard definition of fairness was given by Aristotle in the well-known Chapter 14 of his fifth book on ethics: "The nature of fair lies herein: correct a law that is imperfect due to its universal nature" (Aristotle, 1925). The common law works to restore or restitute the rights. The meaning of fairness in law is to ensure protection in a specific situation where the application of law does not lead to a fair outcome.

The UN International Court of Justice notes that "the justice does not attempt to make equal what the nature made unequal" (ICJ Reports, 1985). If we look at this thought from the opposite side, it appears that fair protection is only justified in artificially created situations or circumstances of inequality. Thus, the "principle of distributive fairness generally means that distribution is fair, if everyone has the right to what they receive as a result of such distribution" (Nozik, 1974). The key point here is "has the right," which means a meritocratic basis of fairness, where everyone has the right to what they deserve according to his or her abilities or what they lawfully acquired previously. Other authors exhibit a different approach, where fairness is based on satisfying the needs. "The principle of need" is based on the opinion that all people have the right to satisfy their certain basic needs. The Marxist motto "to each according to his needs" precisely expresses this position. Fairness in distribution is also induced by the necessity to protect the weak, the poor, and the vulnerable (Konow, 2001).

All these views have influenced the formation of the concept of international environmental fairness. Calls for fairness in the international environmental policies as related to the development of the Arctic's nature and resource potential mainly focus on the need to ensure equal access to the natural wealth and environmental services (Byrne et al., 1998). The development of the Russian Arctic began as early as the second half of the eighteenth century. However, starting from the past decade, in view of the arising environmental problems and through international efforts, the Arctic is now viewed as a single territory with a unique ecosystem rich in natural resources, and a legal framework is gradually being established for its development and protection. So far, this process has been going on as a "soft" law, by adopting political and legal documents for strategic planning, outlining common political approaches and long-term objectives for practical action (Bogolyubov & Krasnova, 2018).

As in law, fairness in the environmental policies was mainly expressed through distribution (Banuri et al., 1996). In the literature, there are different approaches to fair distribution, such as the following:

- "Do not envy" – this expresses the ideal of equal consumption abilities and reflects a situation where no one encroaches upon another's consumer basket (Diamantaras & Thomson, 1990). This also means an equality of expenses and wealth obtained for all (Varian, 1974).
- "Merit-based fairness" stems from the necessity to ensure means of protection proportionate to the degree of unfairness. Such means of protection should not give rise to a new unfairness.

An approach based on universal equality claims that everyone should have the same income, i.e., the lowest 10% of the population must receive 10% of the total income (Stymne & Jackson, 2000).

Meritocracy states that inequality is acceptable, if everyone had equal opportunities during the initial distribution, and disparity is a result of differing commitment and effort made (Konow, 2001).

Minimum standards or an approach based on satisfying the basic needs is applied to the poorest social groups and asserts that no one's income should be lower than a certain minimum level (Stymne & Jackson, 2000).

In addition to the above principles of fairness, it should be noted that fairness, similar to environmental justice, has ethical and causative justification. The rules of fairness attributed to concepts such as "merit-based fairness" and "do not envy" are based on the outcome that is the cause-and-effect rationale. The same applies to the approach based on universal equality and basic needs. Although the arguments for these infer equal access, they talk about a positive outcome rather than the path to it. However, the meritocracy principle ethically places a greater emphasis on the starting positions. If the initial distribution of wealth is fair, any disparity arising subsequent to that will not be viewed as unfair, because it results from the efforts and commitment applied by the given member of the society, who therefore deserves the benefits. Thus, if you judge by merits, it does not matter how fair the resources that allow the rich to ensure their well-being were acquired (Bhaskar, 1995).

Fairness includes the correctional and preventive components. The correctional component provides protections if the common laws fail to produce a fair outcome.

This is similar to the use of fairness in the international and national environmental policies. The concept of "common but differentiated responsibility" in the Convention on Climate Change is designed to correct the obvious disparity in the ability to protect the climate and to avoid the exacerbation of existing unfairness. It also bears in mind future fairness. Since the rules of fairness to be adopted are determined by law or by the distribution of international resources, they become a preventive criterion for identifying the ways in which the problems of fairness will evolve in the future.

It is evident that while environmental justice encompasses and goes beyond concerns about the inclusion of the procedural side of justice in the distributive aspects, fairness is essentially aimed at distribution. Therefore, if environmental justice has both spatial and procedural nature, fairness is mainly associated with addressing the issues of distribution (Robinson, 2002). However, both environmental justice and fairness use ethical and causative arguments. This assumes that in order to establish and develop environmental justice, we have to consider the broad, comprehensive concept of it, which covers all issues of justice in the environmental decision-making process. In this sense, environmental justice will have both the distributive and procedural dimensions. The distributive dimension represents exactly what is usually referred to as fairness and is related to the outcome that people have from social changes (Brashear et al. 2002). The procedural dimension on the other hand is related to procedures and processes (Sheppard et al., 1992). In the broad sense, environmental justice as applicable to the Arctic region can be viewed as preventive, correctional, and punitive justice.

Thus, the authors attempt to clarify the concepts of fairness and environmental justice, which are often mixed into one concept in the environmental literature. For this purpose, the philosophical roots and historical views on these concepts are considered. The concept of environmental justice is proposed as a broad general concept that includes all issues of justice in the process of making management decisions in the field of environmental safety in the Arctic, including the fairness of the procedure and fairness in distribution. On the other hand, when Russia, the United States, Canada, Denmark, and Norway have historical rights to the Arctic sectors, geostrategically subarctic and extra-regional States are also interested in the Arctic, and on the other hand, the world's environment as a whole depends on the state of its environment.

Chapter 2
Geopolitical Risks for Oil and Gas Industry in the Arctic Zone of the Russian Federation

Abstract This charter discusses the key geopolitical factors that affect the sustainable development of the Arctic. Among them are geographic, economic, and military factors. The most important risk factors for oil and gas development in the Arctic are related to gas exploration projects. Analysis of their transformation into opportunities and threats is one of the priority tasks of oil and gas facilities. At the same time, the authors draw attention to the following key geopolitical risks (GPRs): (1) ensuring access to sufficient reserves of hydrocarbon raw materials in the Arctic from various states and obtaining control rights over its natural resources; (2) uncertainty of the legal status of the Arctic region; and (3) geoecological risks (GER) as one of the priorities of attention to Russia's actions in the Arctic. Analysis of GPR in terms of their transformation into opportunities is a priority task of oil and gas facilities in the implementation of Arctic field development projects, especially in the context of raw materials supercycle of energy prices falling in the world.

Keywords Arctic · Oil and gas development · Geopolitical risks · Geoecological risks · Geopolitical challenges and factors · Threats and opportunities

2.1 Introduction

From the point of view of the global geopolitical processes, one of the major factors determining the arrangement and interaction of various geopolitical forces in the twenty-first century is the struggle for resources (Zhiznin & Timokhov, 2019). For this reason, an objective increase in geopolitical contradictions in the Arctic is inevitable. It is connected with its resource potential and importance of transport, on the one hand, and with the absence of a recognized and legally formalized demarcation of sea spaces and the shelf, on the other hand. Furthermore, experts from leading world powers predict the possibility of military conflicts in view of growing contradictions because of division of the colossal wealth of the Arctic (Nuryshev, 2012). Current Russian development of Arctic hydrocarbon resources is associated with

V. N. Bashkin, O. P. Trubitsina, *Geoecological and Geopolitical Risks for the Oil and Gas Industry in the Arctic*, Environmental Pollution 29,
https://doi.org/10.1007/978-3-030-95910-4_2

geopolitical challenges, the essence of which can be interpreted as the emergence of qualitative signs of changes in the evolution of geopolitical factors affecting the processes of sustainable development of the Arctic region (Mitko, 2020).

For oil and gas companies, operating in the Arctic, risk management is a current problem due to specifics of polar conditions. Therefore, top managers of oil and gas companies revise their business strategies, planning, and operation models of risk management (Trubitsina & Bashkin, 2017a) in the Arctic region. In this modern society, this subject is being watched carefully for reasons of geopolitical interest that many countries have due to the natural resource potential of the Arctic. The international community views its vast resources as something that should be shared by the entire humanity so as to meet the energy needs of the planet that is expected to grow in future.

According to a UN report, the world's population continues to grow, albeit at a slower pace than at any time since 1950 (Fig. 2.1).

The world's population reached 7.7 billion in mid-2019 (and already more than 8 billion in the 2021), having added one billion people since 2007 and two billion since 1994. The growth rate of the world's population peaked in 1965–1970, when it was increasing by 2.1 percent per year, on average. Since then, the pace of global population growth has slowed by half, falling below 1.1 percent per year in 2015–2020, and it is projected to continue to slow through the end of this century. However, the global population is expected to reach 8.5 billion in 2030, 9.7 billion in 2050, and 10.9 billion in 2100. Based on a more detailed report of the United

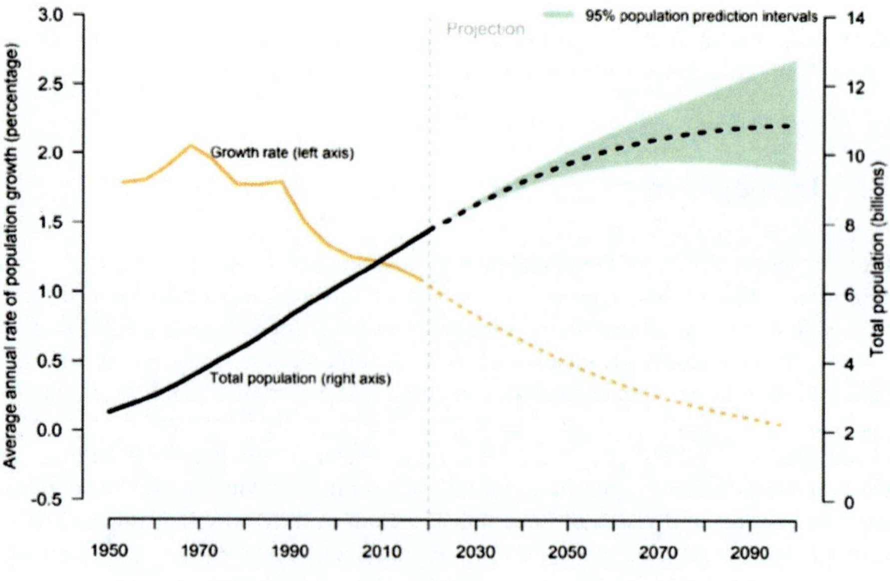

Fig. 2.1 Population size and annual growth rate for the world: estimates, 1950–2020, and medium-variant projection with 95 percent prediction intervals, 2020–2100 (United Nations, 2019)

Nations, with a certainty of 95 percent, the size of the global population will stand between 8.5 and 8.6 billion in 2030, between 9.4 and 10.1 billion in 2050, and between 9.4 and 12.7 billion in 2100. A continuous and rapid population growth presents challenges for sustainable development, putting pressure on already strained resources in the world. Oil, coal, and gas together account for the majority of global primary energy sources. For many countries or regions, including some Small Island Developing States, challenges to achieving sustainable development are compounded by their vulnerability to climate change, climate variability, and sea-level rise (United Nations 2019).

During the study of the main trends in risk ratings for oil and gas companies (Trubitsina & Bashkin 2017b), the role of a position of "geopolitical significance" found out to be among the key points, and this is related to the access to reserves and markets as the limiting factors of a political nature and competition for proven reserves. Owing to this fact, the analysis of geopolitical risks (GPR), especially from the point of their transformation into opportunities and threats, becomes a priority task of oil and gas facilities in the implementation of Arctic field development projects.

Currently, the COVID-19 pandemic is a key challenge in the world apart from population's growth. The global economic crisis is reducing the need for not only oil and gas but also other energy sources. However, energy demand has always been cyclical. Rise and fall in commodity prices can be predicted by identifying commodity cycles. For instance, this is pointed out by José Antonio Ocampo – the UN Under-Secretary-General for Economic and Social Affairs and economist at Columbia University – and his colleague Bilge Erten, based on results of a study of supercycles of rising oil prices (Fig. 2.2). As far as they predicted, markets will approach a cyclical downturn and, therefore, oil prices will fall in 2020 (Commodity cycles, 2013). The forecast has come true. Since that year, a new supercycle of decline has started, and the recovery of global energy resources demand is currently largely connected with the solution of geopolitical contradictions.

In the event of compliance with all international sources of law, oil and gas companies could focus more on Geoecological risk (GER) issues, which occupy a dominant place in the risk ratings among such companies (Trubitsina & Bashkin, 2017b). However, in fact, the situation is not the same. For example, the United States often does not consider Russian internal decisions on maritime borders, particularly, in the Sea of Japan and in the Arctic Ocean (AO). At the same time, China is not yet moving toward decarbonization and is very interested in the Liquified Natural Gas (LNG) transport by North Sea Route (NSR) and in the production of which they have already invested a lot of money (Arctic-LNG-2). Large reserves of oil and gas have been discovered in the area of the Novosibirsk Islands, the production of which, as well as at the Shtokman oil and gas condensate field, will begin when it is economically profitable. Since the Russian Federation intends to defend this region, including by military means, this indicates the presence of threats, including those discussed in this chapter. It is aimed at identifying key geopolitical factors affecting the sustainable development of the Arctic, as well as analyzing the similarities and differences in the geostrategic positions of the Arctic Five states. At the same time, the authors focus on positions of the GPR, associated with ensuring access and

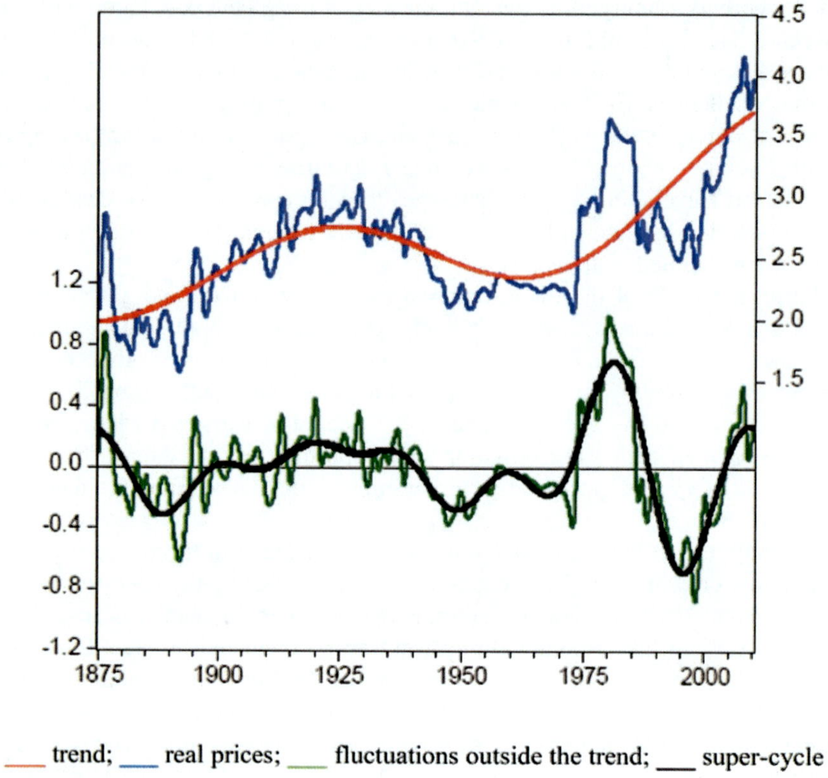

Fig. 2.2 Oil price dynamics (Ocampo, 2013)

obtaining control rights over the hydrocarbon resources of the Arctic from different countries, the uncertainty of legal status of the Arctic region, as well as the use of geoecological risks (GER) as a manipulation tool to draw attention to Russia's actions in the Arctic.

2.2 Geopolitical Challenges and Factors for Russia in the Arctic

The geopolitics of the Arctic as a macro-region is determined by its position in relation to other countries in terms of similar or different positions of political systems and geopolitical potentials in conjunction with the presence or absence of mutual interests and problems (Baklanov et al., 2015).

Historically, the Arctic has not been under geopolitical pressure due to its harsh climatic conditions, insignificant and low population density, and geographical location on the periphery of world events. Nevertheless, during the Second World War, the great importance of the Arctic region against the backdrop of renewed

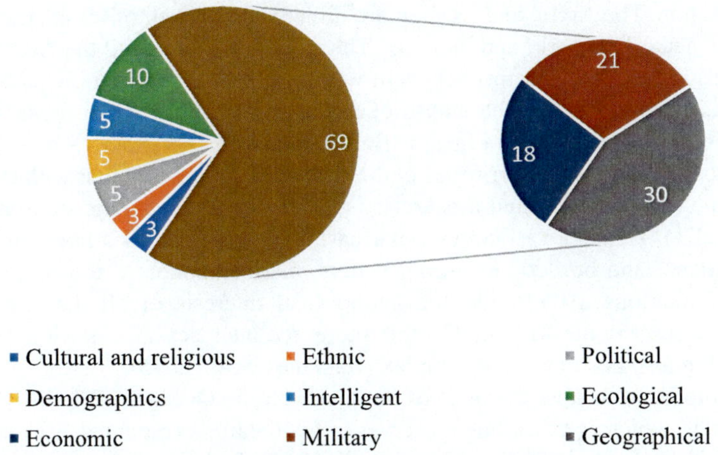

Fig. 2.3 The contribution of geopolitical factors to the sustainable development of the Arctic, %, according to materials (Baklanov et al., 2015)

conflict between the great Powers and the emergence of modern transport technologies and global commodity supply chains was demonstrated. In North America, both in Canada, and the United States, transport links in their polar territories and surveillance of them was strengthened to prevent Japanese encroachments on the Aleutian Islands adjacent to Alaska. At the same time, Greenland and Iceland were key transit countries for Allied troops, aircraft, and cargo transportation to Europe. For the Soviet Union, the NSR became a link in the chain of allied supplies of food and equipment to the port of Murmansk and served as a fundamentally important "road of life" in an effort to prevent the invasion of Nazi Germany and its Allies (Pickford & Collins, 2016; Melas, 2016).

In modern time, the Russian mission in the Arctic is determined by geopolitical factors, the evolution of which presupposes both tendencies to increase their influence on sustainable development processes and the redistribution of their share.

The contribution of geopolitical factors to the sustainable development of the Arctic and the relationship with social categories according to expert estimates by scientists from the Arctic Public Academy of Sciences (APAS) is shown in Fig. 2.3. Three key factors have the largest share (69%): (1) geographical (30%), (2) military (21%), and (3) economic (18%).

Challenges related to the geographic factor, spatial location, and natural resources are considered basic in Russia's current context. Taking into account the changes in its territory and approaches to determining the external boundaries of the continental shelf in the Arctic (instead of the sectoral one to comply with the Convention on International Maritime Law), the evolution of the geographical factor was quite significant in the last century. Because of the last phase of geographic changes, Russia has undergone a significant "northernization" in the twenty-first century (Mitko, 2016). Norway and Denmark adjoin Russia through land and sea borders within the Western macro-region and the United States and Canada within the eastern region. The Arctic and its shelf are directly connected to Russia, the United States, Canada, Denmark, and Norway. The reason for obtaining the Arctic status and securing the Arctic sectors for them was their northern borders, which extend beyond the Arctic Circle. The length of the coastline of the Arctic Five states in descending order is shown in Fig. 2.4 (Baklanov et al., 2015).

The coastal and insular territories of the Arctic Five states together with the water area of the marginal seas and the Arctic Ocean make up the Arctic transboundary region (ATR). This is a vast circumpolar basin zone, crossed by a large number of state borders: land borders, territorial waters, marine economic zones, the Arctic shelf delimitations. At present, the geopolitical interests of all these countries already intersect in the ATR, and in the future, the intersection zones will not only increase but also become more complex (Baklanov et al., 2015).

It is important to note that half of the entire Arctic Ocean shelf is the Siberian Arctic shelf, which contains huge reserves of hydrocarbon resources. Special attention should be paid to the East Siberian Shelf (ESS) due to the prerequisites for development of the most serious consequences associated with modern climatic changes. The ESS is the largest and shallowest continental shelf in the World Ocean. With an average depth of about 50 m, it occupies 2.1×10^6 km^2 and covers the Laptev Sea, the East Siberian Sea, and the Russian part of the Chukchi Sea. The entire area of the ESS is covered with underwater permafrost, which in the past 30 years has been degrading at a double rate, freeing up access to marine energy reserves, as well as contributing to methane emissions (Gershelis et al., 2020; Grinko et al., 2020).

The challenges of the economic factor include a decrease in the share of added value of high-tech and science-intensive sectors of the economy in the gross regional product of the Russian Arctic, weak interaction of the research and development sector with the real sector of the economy, and the discontinuity of the innovation cycle (Decree of the President, 2020).

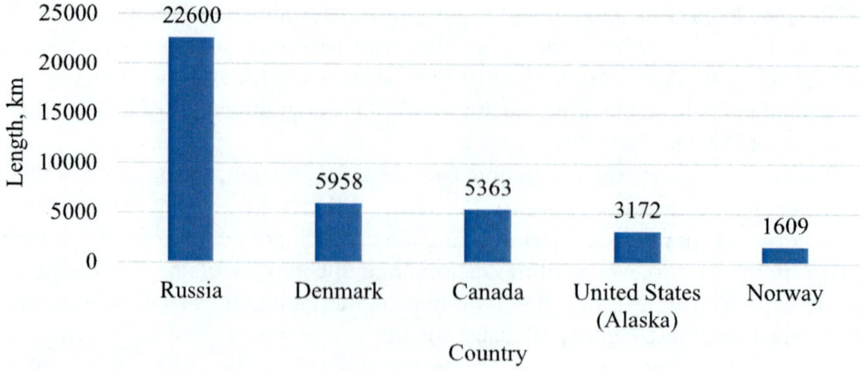

Fig. 2.4 The length of the mainland coast of the Arctic Five states beyond the Arctic Circle (km) according to the materials (Baklanov et al., 2015)

The specific gradient of the military factor evolution in the Russian Arctic has comprehensively influenced the Arctic activities. It is important to note that the military factor is associated with almost all other factors of sustainable development or security of the Arctic society. The evolution of the military factor technically caused an increase in its share in the system of factors and a qualitative transformation of its content, with an emphasis on the main areas, requiring the abandonment of traditional methods of military operations due to environmental, political, humanitarian reasons, and the development of such vectors as information confrontation in the form of "network-centric" strategies, the massive use of non-lethal weapons in the fight against terrorism in the Arctic, and the massive use of robotics (Mitko, 2016). At the same time, the Strategy for the Development of the Arctic Zone of the Russian Federation (AZRF) and Ensuring National Security for the Period up to 2035 indicates the challenge of increasing the conflict potential in the Arctic region, dictating a continuous increase in the combat capabilities of the groupings of troops (forces) of the Armed Forces of the Russian Federation, other troops, military formations, and bodies (Ocampo, 2013).

The authors believe that the share of the environmental factor is underestimated. Owing to the fact that environmental problems in the Arctic identify global trends, it is unacceptable to consider them only as national or regional ones. At present, the desire to make a profit dominates in the Arctic geopolitics, and the current trend of de-ecologicalization not only of Russia but also of the whole world is reflected (Lukin, 2011).

Apparently, climate warming is most evident in the Arctic, as evidenced by a significant increase in air temperature, increased river flow, reduced area of ice cover (Grinko et al., 2020), which certainly requires enhanced environmental monitoring and accounting when making management decisions. In this regard, the issue of increasing the specific weight of the environmental factor is relevant. The transition to sustainable development makes it necessary to include it in the system of basic socioeconomic development indicators. The underestimation of the environmental factor in decision-making is largely due to the lack of value reflection of natural capital and environmental degradation in traditional development indicators. The traditional macroeconomic indicators (GDP, per capita income, etc.) ignore environmental degradation. The growth of these indicators is based on technogenic nature-intensive development, thereby creating the possibility of sharp deterioration in economic indicators in future in case of natural resources depletion and environmental pollution (Yashalova & Ruban, 2014). For example, in the study of determining the relevant indicators for compiling the index of environmental safety of the Russian Arctic and ranking (compiling a rating) of the regions of the Russian Arctic, it is indicated that in the Krasnoyarsk Krai (an outsider of the rating), despite a number of environmental problems, there is a very high level within the regional GDP (Bobylev et al., 2020), which is reflected in the ecological perception of people living there.

2.3 Geostrategic Challenges in the Arctic

In terms of their geostrategic relationship to the Arctic, states can be divided into three groups (Table 2.1), which "compete both among themselves and in the format of international organizations" (Smirnov, 2016; Trubitsina & Bashkin, 2017a).

The first group contains the Arctic Five. It includes the coastal Arctic states: Russia, Canada, the United States, Denmark, and Norway. At first, these five countries met in 1973. As a result, they concluded an agreement on the conservation of polar bears. It is important to note that only in the last 15 years this informal association has developed great activity. It is connected with the influence of climate change, the increased economic interest of non-Arctic states in this region, and friction over claims to new borders in the mid-2000s. At the same time, "Arctic regional governance is best viewed as a web with the Arctic Council in the middle, not a pyramid with the Arctic Council at the top" (Exner-Pirot, 2016). It is known that there are three cases of negotiations where the Arctic Five states have presented a united front: in Ilulissat, in Chelsea (Canada, 2010), and in Oslo (Norway, 2015). Many people point out that the main advantage of the informal structure of the "Arctic Five" is developing binding agreements with non-Arctic countries and making "concrete decisions on issues affecting the interests of different states first of all in the field of security" (Kuersten, 2016).

It is noteworthy that in the first group, the United States, Denmark, Canada, and Norway are (North Atlantic Treaty Organization) NATO members; this exacerbates the potential for a military conflict in the Arctic between NATO and the Russian Federation. Canada, Russia, the United States, and Norway expressed their intentions to develop the Arctic region in the state policy documents, some of the provisions of which coincide in the following positions (Komleva, 2011):

1. Strategic importance of the Arctic region both for the state and for the whole world

Table 2.1 Geostrategic attitude of groups of states to the Arctic

State group number			
The first group		The second group	The third group
States	The Arctic Five states (Russia, the United States, Denmark, Canada, Norway) have access to the Arctic Ocean	Subarctic states (Iceland, Finland, and Sweden) do not have access to the Arctic Ocean but are members of the Arctic Council	Non-regional states (Brazil, India, China, Singapore, South Korea, Japan, EU countries, etc.)
Characteristic	They have the right to develop natural resources of the shelf, the expansion of which to the north is the subject of unresolved interstate contradictions	They do not have rights to the shelf, but they strive to increase their status and influence in the format of the Arctic Council	They try to maximize their geostrategic attitude to the Arctic, influence the revision of its status, referring it to the common heritage of mankind

2. Leadership in the Arctic and implementation of the task of strengthening its sovereignty over the relevant sector of the Arctic
3. Development of the economy and social sphere, environmental protection, scientific research, improvement of the management structure of their own Arctic sector in a circumpolar dialogue regime
4. Military presence as an integral part of its presence in the region: creation of Arctic groups of forces (land and sea), new bases for such groupings, strengthening of border formations, improvement of infrastructure

Along with general positions, there are those that distinguish each state of the first group from others in the form of specific strategies, namely the following:

- The Danish Arctic Strategy, adopted in May 2011 for the period 2011–2020, is based on the Ilulissat Declaration of May 28, 2008, in which scientific, geological data and international law form the basis for future land allocation. This declaration informs the non-Arctic states about the internal nature of issues related to the division of the Arctic and their belonging only to the Arctic countries. It is also noted that a format close to the will not be considered. The 2011 Danish Arctic strategy showed the first noticeable signs of a national aspiration toward the Arctic as opposed to only a narrowly focused view of Greenland earlier.
- The attitude of the Arctic states toward this region has been transformed in conjunction with the Ilulissat Declaration. The assessment of the Arctic importance has become deeper among the states of the Arctic Council (Canada, Denmark, Finland, Iceland, Norway, Russia, Sweden, and the United States), founded on September 19, 1996, which was reflected in the formation of guidelines for the foreign and domestic policy of the Arctic Eight. The chronological list of states, that have formulated their Arctic strategy is as follows: Norway (2006), Russia (2008), Canada (2009), Finland (2010), Iceland (March 2011), Sweden and Denmark (May 2011), and the United States (2013) (Allayarov & Shubin, 2017). Orienting to the activities of a number of environmental working groups established in the early 1990s, the Arctic Council focuses on environmental protection and sustainable development. Neither issues of security and defense nor trade and immigration are discussed by the Council. Decision-making is based on the consensus or unanimous consent of all eight countries. As a result of this, the disparity of forces between these countries militarily, economically, and geographically is considered. Consensus decision-making is the key to ensuring that "the Council will not be used to impose a certain policy" on any one state (Bloom, 1999).
- Fundamentals of the state policy of the Russian Federation (RF) in the Arctic for the period up to 2020 and beyond were approved by the President of the Russian Federation on September 18, 2008. The main national interests of Russia in the Arctic contain use of the Arctic zone as a strategic resource base that provides social and economic development to the country, preservation of the Arctic as a zone of peace and cooperation, preservation of unique Arctic ecosystems, and

use of the Northern Sea Route (NSR) as a national unified transport communication of Russia in the Arctic.

- Decree of the President of the Russian Federation of October 26, 2020, No. 645 approved the "Strategy for the Development of the Arctic Zone of the Russian Federation (AZRF) and Ensuring National Security for the Period up to 2035" (Decree of the President, 2020) in order to ensure the national interests of the Russian Federation in the Arctic zone, as well as to achieve the goals defined in the Fundamentals of the state policy in the Arctic. There are also detailed measures for the three-stage implementation (2020–2024, 2025–2030, 2031–2035) of the main tasks in the spheres of social, economic, infrastructural, scientific, technological, environmental development and international relations, ensuring both military security and safety from natural and anthropogenic emergencies. The new Strategy has a special regional section that defines the main directions for the implementation of the Strategy for each territory within the AZRF.
- Norway's strategy for the Arctic is different from the Arctic states in the desire to develop the region in an ideological space, along with geographic and economic. This approach is reflected in Norwegian-Russian relations, recognized in the text of Norway's Northern Strategy. Therefore, for the implementation of learning and research processes in educational institutions of Northern Norway, the Scholarships of the Northern Regions are established for students and scientists from Russia. In this way, there is a certain degree of consciousness transformation of the fellows, aimed at implementing the policy of the "country of study" within the geopolitical spaces of other societies that are native to the fellows. Norway, actively preparing to the struggle for its interests in the Arctic, uses the so-called "soft" power, not excluding the development of "hard" power (Komleva, 2011).
- The Canadian Arctic Strategy "Canada's Northern Strategy: Our North, Our Heritage, Our Future" (Canada's 2009) focuses on aspects of public policy related to the integrated development of the northern territories. The document highlights the position emphasizing that the north is an integral part of the identity of modern Canada, historically formed even before the arrival of Europeans to the American continent and associated with the continued development of the north by indigenous peoples. This position is supported by the majority of Canadians, who consider the confirmation of the rights to the Arctic as a priority of the foreign policy of modern Canada (Konyshev & Sergunin, 2011). As an Arctic country, Canada claims an active leadership role in shaping the governance, sustainable development, and environmental protection of the strategic Arctic region, as well as interacting with other countries to advance its interests (Statement, 2010).
- The US Arctic Policy Directive of January 12, 2009, emphasizes that "the United States has broad fundamental national security interests in the Arctic and is prepared to act independently or in alliance with other states to protect these interests" (National Security, 2009; Konyshev & Sergunin, 2011). The US government's strategic priorities in the Arctic are reflected in the US National Strategy for the Arctic Region. For example, the security sphere includes anti-

missile defense and warning, deployment of maritime and air systems for strategic maritime transportation and strategic deterrence, and operations to ensure maritime security and freedom of the seas, including the NSR. In general, the strategy focuses on ensuring the country's security interests, responsible management in the context of protecting the Arctic environment, and preserving its resources and developing international cooperation in the Arctic (US National Strategy for the Arctic, 2013). Innovations regarding the US government planning for the Arctic region were announced on June 6, 2019, in the Arctic Strategy of the US Department of Defense, which updated the previous 2016 strategy. The new document contains a secret appendix and context of the rivalry of different countries, security threats from Russia and China, highlighted by Secretary of State Michael Pompeo in Finland (Looking North, 2019). Previously, the Arctic Council had hardly discussed security issues, mainly addressing climate change, environmental protection, and sustainable development in the region. In this regard, since 2019, there is a new tendency to take security issues into account in the context of national rivalries (Gorobets, 2019).

As for states of third group, one of the main challenges facing the circumpolar states is how to consider the interests of non-Arctic countries in the region without jeopardizing their own dominance in Arctic politics? Particularly, Asian countries and the European Union express a desire to become more active in the region. These States strive to have a strong say in solving Arctic problems, to which they are pushed by a combination of environmental, economic, and trade factors (Collins, 2017a, b).

2.4 Threats and Opportunities of GPR for Hydrocarbon Development in the Arctic

The Arctic has huge oil and gas reserves and is thought to include about a quarter of the world's undiscovered oil reserves: most of them are situated in Alaska, northern Canada, Norway, and Russia, considering significant amounts in offshore areas. Continuing reduction of sea ice is likely to result in increased oil and gas activity on the shelf, particularly in terms of increased offshore oil transportation as the navigation season lengthens and new sea routes open (Bashkin et al., 2016). Nevertheless, warming in the Arctic has an opposite side, which is the gradual destruction of the Polar infrastructure, created in permafrost conditions (Mitko, 2020).

It is believed that comprehensive attention to hydrocarbon projects of the Arctic shelf is based on the likelihood of discovering the largest deposits here. It is connected with onshore discoveries in the last decade and is characterized by small reserves. Easily accessible oil and gas resources have already been discovered and used. It is predicted that fossil fuels will remain a significant source of energy until 2050. At the same time, according to prognosis, global energy demand will have grown by more than a third by 2035 alone. As the owner of one-third of the world's known natural gas reserves and the largest oil-producing country in the world, Russia is interested in the Arctic as an area of new opportunities, despite geopolitical and geoecological challenges. This region will play an important role in satisfaction of the global energy supply in the next few decades (Trubitsina & Bashkin, 2017a).

In the event of Arctic exploration, it is necessary to take into account the environmental factor, and due to the expansion of oil and gas development projects, particularly on the sea shelf, the ecology of the region may suffer. Referencing to the Strategic Action Program for Environmental Protection in the Russian Arctic,

> "... the increased rates of the oil and gas industry development in the Arctic zone of the Russian Federation (AZRF) in the last decade and the planned development of work on the shelf of the Barents Sea and other Arctic seas create a threat of escalation of the local scale of environmental degradation into a zone-wide one. At present, the direct flow of crude oil into the marine environment, freshwater reservoirs, and landscapes of the coastal areas of the Russian Arctic is of a limited nature and is not considered as a factor that significantly complicates the general zonal ecological situation. The danger of pollution of the marine environment by oil is associated with plans for its production on the continental shelf of the Russian Federation" (Trubitsina, 2016a).

Activity in the oil and gas industry of the Russian Arctic has been growing in the past few years, and the GER is also growing accordingly. This leads to the formation of "hot spots" and "impact zones," characterized by a high level of chemical pollution of the environment and transformation of the natural geochemical background; degradation of marine flora, vegetation, soils; uncontrolled development of erosion; cryogenesis; formation of sinkholes in vast areas; influx of pollutants in the food chain; a high level of morbidity in the population; air pollution with strontium compounds, heavy metals (in particular mercury), oil products, etc. (Morgunov, 2011; Trubitsina & Bashkin, 2016).

For the above-mentioned reasons, it is extremely important to strengthen the significance of the ecological position in the common structure of geopolitical factors affecting the sustainable development of the Arctic region. At present, many experts in the world are studying the assessment of probability of environmental hazard in the absence of an unequivocal answer about the impact of chemical pollution of modern industries on natural ecosystems. Nevertheless, despite the type and nature of production, an enterprise is an element that determines the structural relationship between it and natural environment. In fact, such interconnection in Arctic is very special due to the fragile nature of high latitudes that extremely vulnerable to anthropogenic impact and mutual affection of harsh arctic climate conditions on technical components of modern oil and gas enterprises (Trubitsina, 2015).

Overall, the GPR is caused by global processes and trends in use of natural resource potential of the Arctic, both in the interests of the world and individual countries. Possible manifestations of the GPR are the violation of the system of strategic stability in the Arctic geostrategic space. Moreover, the GPR represents the likelihood of a change in the geopolitical situation at the regional and global levels, expressed in unfavorable conditions (risk of a hybrid war, military clashes, etc.) or additional opportunities.

The issues of the NSR have contradictory legal subtleties. Norms of the UN Convention on the Law of the Sea of 1982 (hereinafter the Convention) are on the side of unrestricted exploitation of this highway by ships of foreign states. According to the Convention, ships of any states have the right to free navigation within the exclusive economic zone of the coastal state, which is equal to the distance of 200 nautical miles from the coastline minus 12 nautical miles of the territorial sea and 12 miles of the adjacent zone. Taking into account this rule, part of the NSR can indeed be freely used by foreign vessels, military ones among others. Despite this fact, the difficulty of the situation is that the sea route along the northern borders of Russia is very changeable, and its configuration depends on freezing of the seas, weather, and hydrological conditions.

In response to foreign claims, the Russian leadership not only declares the northern transport highway "a historically established national transport communication" but it is also referring to Article 234 "Ice-covered areas" of the Convention. In an extreme climate and severe ice conditions, coastal countries (in this case, Russia) can independently regulate shipping to prevent possible environmental damage. Coastal countries are responsible for Safety and Disaster Prevention. Borders are legally described as "ice-covered areas." Moreover, according to the tradition that has developed over the centuries, the powers over the territory are transferred to the pioneer state. From this point, Russia has much more legal "bonuses."

However, both the United States and the northern countries of the NATO bloc are not entirely satisfied by forming situation. NATO warships are increasingly appearing in the Arctic region, guided by a one-sided understanding of the provisions of the Convention. Potential objects of control not only are the NSR itself but also plentiful deposits of minerals hidden in the continental shelf. Trainings are being organized in the neutral waters of the Arctic zone, in which not only NATO countries participate but also "neutral" Sweden and Finland.

According to the Russian defense ministry in 2019, the intensity and scale of operational and combat training activities of NATO's armed forces in the Arctic have increased by 17%, while intelligence activities have increased by 15%. Missile defense systems are being strengthened in coastal states. Some countries are resuming underwater patrols in the region.

Since 2018, the second operational fleet of the US Navy has been reestablished and its area of responsibility included part of the NSR off the Russian coast. At the same time by 2022, the United States together with the Europe are going to form a joint NATO command "Atlantic," which, together with the support command, will ensure the rapid transfer of American troops to Europe. Moreover, at a distance of 60 km from the Russian border, the Norwegians are building a new radar station. Besides this, Norway doubled the number of US Marines deployed in 2018. British submarines with "Tomahawks" have been patrolling the Arctic since 2016, while ground units are honing their "war in the cold" skills in Norway (Fedorov, 2020).

Thus, taking everything into account, the main threats and opportunities of the GPR are as follows:

1. ***Ensuring access to sufficient reserves of hydrocarbon raw materials in the Arctic from various states, obtaining control rights over its natural resources.***

   Threats:

   1. Depletion of traditional hydrocarbon deposits, for instance, depletion of "light" oil and low-permeability gases resources
   2. Necessity to search for new oil and gas sources and transfer of exploration to more inaccessible areas
   3. Loss of control over the Arctic territories
   4. Military confrontation of the polar countries on issues related to the delimitation of the Arctic shelf and oil and gas resources located on it

   Opportunities:

   1. Increased development of unconventional hard-to-recover deposits
   2. Expansion of the resource base, including by increasing the share of oil and gas production in the Arctic regions with difficult conditions and low development
   3. Development of advanced technologies to exploit new Arctic reserves, previously considered unprofitable due to difficult natural and climatic conditions
   4. Ensuring stable access to hydrocarbon reserves
   5. Resolving controversial issues of the Arctic territories ownership by global consensus or consensus of global policy actors
   6. International cooperation with the attraction of foreign investments and technologies while maintaining the national interests of the state
   7. Development of necessary technologies and resources to reduce the level of GPR

2. ***Uncertainty of the legal status of the Arctic region***

   Threats:

   1. Increase and complication of the current position of the geopolitical interests' intersection zone of the main geostrategic and regional players
   2. Uncertainty in the interpretation of unified international requirements and mechanisms for their application

   Opportunities:

   1. Resolving controversial issues of the Arctic territories ownership by global consensus or consensus of global policy actors
   2. Unification of regulatory requirements and creation of a unified international mechanism for regulating the companies' activities in the Arctic
   3. GER as one of the priorities of attention to Russia's actions in the Arctic

   Threats:

   1. Putting pressure on Russia in the context of its plans to develop Arctic infrastructure and build an oil and gas complex. The goals and actions of the Arctic states are aimed at proving lack of legal grounds for Russia to develop offshore fields, to use the NSR as an internal passageway, and to blame

Russia for its inability to ensure environmental safety when developing deposits in the region.
2. Threat of Russian "hybrid war" in the Arctic in the format of coordinated use of political-diplomatic, information-psychological, economic, and power tools to achieve strategic goals. In NATO expert circles, as a rule, the concept of "hybrid wars" is already used to denote the role of Russia in crisis points.
3. Manipulation of GER through geopolitical provocations in the context of inability of Russian oil and gas facilities to ensure environmental safety in the Arctic. For example, Greenpeace activists advocate the Arctic development as a whole, but oppose individual projects that damage the region's ecology. At the same time, the danger of the project is determined by Greenpeace itself. As a rule, Russian projects (Gazprom and Rosneft) regularly find themselves among the environmentally hazardous ones.
4. The sanctions policy against Russia by the European Union and the United States is also aimed at weakening influence in the Arctic region.
5. In connection with the above point 4, low oil prices are also perceived by foreign initiators as one of the factors limiting Russia's resources in the NSR development.

Opportunities:

1. Russia has developed special rules for the passage of foreign military vessels along the NSR as a retaliatory measure:
   - Firstly, notification of a warship visit must be delivered at least 45 days in advance. The document must reflect the ship's name, sailing time, and a clear route. Moreover, the notice separately describes the vessel's displacement, draft, and propulsion parameters. Formally, all this is required by Article 234 "Ice-covered areas" of the UN Convention.
   - Secondly, a Russian marine pilot is mandatorily sent to a military vessel. The movement of the vessel is under the full control of the Russian Navy. In case of an emergency, icebreakers will come to the aid of a military vessel.
2. Development of necessary technologies and resources to reduce the level of GPR and reduce the abovementioned threats.

2.5 Conclusion

Modern human activities in the Arctic region should be aimed at the support of sustainable development of the polar territories. The rational placement of production industries, in particular, oil and gas, which meet the protection and restoration of the environment in the changing climatic and geopolitical conditions contributes to this.

The key geopolitical factors affecting the sustainable development of the Arctic are geographic, economic, and military. It is necessary to strengthen the role of the ecological factor. Furthermore, the environmental problems of the Arctic are an indicator of global trends, and they cannot be regarded as purely national or regional. Despite geopolitical constraints, the strategic importance of the Arctic is growing. International political, military, and legal disputes over the possession of its territories, connected with economic interests, are intensifying.

The most important risks for oil and gas development in the Arctic are gas exploration projects; analysis of their transformation into opportunities and threats is one of the priority tasks of oil and gas facilities. At the same time, the authors draw attention to the following key GPRs: (1) ensuring access to sufficient reserves of hydrocarbon raw materials in the Arctic from various states and obtaining control rights over its natural resources; (2) the uncertainty of the legal status of the Arctic region; and (3) GER as one of the priorities of attention to Russia's actions in the Arctic. Analysis of GPR in terms of their transformation into opportunities is a priority task of oil and gas facilities in the implementation of Arctic field development projects, especially in the context of the super cycle of energy prices falling in the world.

Chapter 3
Geoecological Risks for Oil and Gas Industry in the Arctic Zone of the Russian Federation (Analyses and Modeling)

Abstract The use of methods for calculating geoecological risks and for assessing the probability of their excess in the impact zones of oil and gas projects in the Arctic region has been proceeded. The analysis of data and methodological approaches to the assessment of geoecological risks in connection with the anthropogenic impact of projected liquefied natural gas (LNG) production and natural gas production in the shelf zone of the Arctic seas revealed the differentiated nature of the stability of various aquatic ecosystems in relation to atmospheric emissions of nitrogen oxides. Furthermore, the impact of existing gas production and transportation projects on the probability of geoecological risks was carried out in the coastal Arctic regions such as Taz Peninsula and Yamal Peninsula. These risks largely depend on the intensity of the work of the gas treatments plants and, accordingly, the probability of geoecological risks is higher in impact zones adjacent to point sources of NOx emissions. In some cases, due to the effects of vegetation change, there is a high probability of risks of soil thawing, which can lead to a violation of the stability of piles of gas pipelines and foundations of engineering structures and road facilities. This, in turn, requires the reclamation of such ecosystems.

Keywords Geoecological risk · Arctic marine ecosystem · Arctic coastal ecosystem · Impact zone · Modeling

Already at present, the main explored and operational reserves of energy carriers, both gas and oil, are concentrated in the Arctic region of Russia. Gas resources of coastal areas are mainly in operation. In the Arctic zones, these are gas fields on the Taz Peninsula and the Yamal Peninsula. At the same time, oil production has already begun on the Prirazlomnaya platform in the Pechora Sea, and gas production is planned in the Ob and Taz bays. It is also necessary to note the promising gas production works at the giant Shtokman gas field in the Barents Sea. In addition, given the changing climate and the appearance of icebreaker-class tankers, there is a

This chapter was written in cooperation with Irina V. Priputina, PhD (Institute of Physico-Chemical and Biological Problems of Soil Science of the Russian Academy of Sciences)

V. N. Bashkin, O. P. Trubitsina, *Geoecological and Geopolitical Risks for the Oil and Gas Industry in the Arctic*, Environmental Pollution 29,
https://doi.org/10.1007/978-3-030-95910-4_3

prospect of year-round use of the Northern Sea Route for liquefied natural gas (LNG) transportation, which was demonstrated in 2020. However, already in 2021, the ice situation in the Arctic Ocean was not so optimal, and such changing weather conditions impose certain restrictions on the development of oil and gas projects and the transport of hydrocarbons.

Therefore, it is necessary to consider the geoecological risks arising from the development of oil and gas fields in the Arctic region, both in marine and coastal zones.

3.1 Assessment of Geoecological Risks in Impact Zones of Arctic Marine Ecosystems

The implementation of oil and gas projects in the Arctic seas requires a qualitative and quantitative assessment of geoecological risks. Such projects, in particular, at gas condensate fields, including the extraction of natural gas from offshore fields, its transportation to processing and transportation sites, as well as liquefaction at LNG plants with subsequent delivery by sea to consumers, are already included in the current and long-term plans of many oil and gas companies. Taking into account the specifics of the Arctic, this determines the need for a comprehensive scientific assessment of all aspects of the implementation of these projects, including their environmental component (Howarth, 1996; Romankevich & Vetrov, 2001; Moiseenko, 2005; Dushkova & Evseev, 2011; Willemse & Van Gelder, 2011; Arctic Standards Recommendations, 2013; Garmo et al., 2014; Moiseenko et al., 2016; Alekseeva et al., 2019). Among the basic principles we should note: compliance with the requirements of international legislation in the field of marine environment protection; development of environmental requirements for each stage of LNG projects implementation; creation of sectoral environmental standards; development of scientific and methodological documentation for LNG project support and, as a result, assessment of geoecological risks (Bashkin & Priputina, 2010; Meshcherin et al., 2011). Unfortunately, the current experience in implementing such projects is limited and does not cover all issues, especially those related to their environmental safety. The increased uncertainty of the environmental consequences of gas production in the offshore zone of the Russian Arctic is determined by (1) insufficient information on the stability of marine ecosystems of the Far North to the specifics of anthropogenic impacts related to gas industry and (2) the diversity of natural-territorial complexes (NTCs) involved in the zone of impact of gas production and processing and transportation facilities.

Accordingly, a scientific and practical study of the issues of qualitative and quantitative assessment of geoecological risks arising from the environmental impact of gas production and LNG production facilities, as well as the impact of the changing natural environment on these production facilities, is required. In the general structure of the gas industry, LNG production is included in the gas processing subsystem, for which the main geoecological risks are environmental pollution, impact on human health, socio-environmental risks, risks of construction, and operation of gas-processing facilities in difficult climatic conditions. The use of hydrocarbon raw materials from offshore fields for LNG production and its tanker transportation to the consumer lead to the fact that the combined impact zone of all technological and related facilities includes coastal territories and marine waters, between which there are special interactions of matter and energy flows, which further complicates the assessment of geoecological risks.

As is known, there are no strict methods of accounting for geoecological risks to the natural environment at the ecosystem level yet. The main attention of marine ecologists, as a rule, is paid to the prevention of the consequences of oil spills and pollution of seafood consumed by humans with toxic compounds coming from coastal drains (http://www.imo.com). There are some publications in the literature regarding the effects of pollution of marine ecosystems due to the deposition of atmospheric pollutants on the surface of the water area (Bashkin et al., 1995; Golubeva, 2007; Willemse & Van Gelder, 2011; Arctic Standards Recommendations, 2013; Alekseeva et al. 2019; Ma, 2020). Earlier, pilot assessments of environmental risks for terrestrial ecosystems because of nitrogen oxide emissions were carried out for a number of gas industry facilities, which showed the effectiveness of using the basic principles of the risk analysis methodology for different levels of NTC (Bashkin et al. 2002a; Samsonov et al. 2007a; Bashkin, 2014; Bashkin & Kazak, 2015). However, for aquatic ecosystems, there are no scientific and methodological developments and data necessary to establish causal relationships between the intensity of anthropogenic impacts and the responses of marine biota communities. Under these conditions, for probabilistic modeling of events associated with possible risks to the environment, the most promising is the use of a systematic approach involving expert statistical assessments.

Using the example of the LNG plant planned by Gazprom in the future, which is located in the northern part of the Kola Peninsula on the coast of the Barents Sea and based on the raw materials of the Shtokman field (Fig. 3.1), a possible algorithm for assessing geoecological risks associated with LNG production and transport is considered.

The proven reserves of the Shtokman field amount to 3.8 trillion m^3 of gas and more than 31 million tons of condensate. With reaching the design capacity, about 45 million tons of LNG can be produced for the raw materials of the field (Meshcherin et al., 2011).

Due to the complex nature of the interactions of modern production systems with the natural environment, the practical solution of the problems of assessing and managing these risks is carried out through the decomposition of a multifactorial complex of interactions into separate components described using relatively simple models and the subsequent synthesis of the results obtained to search for strategic solutions (Samsonov et al. 2007a). As a leading risk factor in this study, an increase in regional levels of nitrogen oxides emissions into the atmosphere as a result of the combustion of process gas during the development of an offshore field and the operation of an LNG plant is considered. This factor is included in the subcategory of territorial risk factors, determining the negative impact of gas industry facilities on

Fig. 3.1 Location of research objects

the environment and the dependence of risk indicators on natural and climatic conditions (Rusakova et al., 2009). The location of the projected objects in the shelf zone of the Barents Sea and the coastal zone of the northern part of the Kola Peninsula determines the diversity of recipient groups, including the following ecosystems:

- The shelf zone of the Barents Sea in the area of the location of gas production facilities of the Shtokman field
- Coastal areas of the Barents Sea near the location of the LNG plant
- Terrestrial ecosystems within the tundra and forest tundra zones of the Kola Peninsula in the area of the LNG plant location
- Freshwater ecosystems of the northern coast of the Kola Peninsula in the area of the LNG plant location

It seems optimal to use the standard scheme of the risk assessment algorithm (Risk Analysis, 1988; Bashkin, 2006a, b), when a set of sequential studies is performed for each of the groups of recipients under consideration:

1. Identification of potential hazards associated with the transformation/disruption of the structure and functioning of NTC or their edifying components because of pollutants entering the environment
2. Assessment of the recipients'exposure under different scenarios of the design capacity of technological facilities

3. Assessment of the dose-response relationship characterizing the permissible level/dose of exposure to pollutants in relation to certain effects of disturbances in the structure and functioning of ecosystems

Risk characterization is an assessment of the probability of adverse environmental effects based on the calculation of excess permissible parameters of impacts.

In accordance with the general provisions of the risk methodology, when performing quantitative assessments of environmental risks, scenarios of the most adverse consequences and maximum impacts are traditionally considered.

3.1.1 Assessment of Environmental Risks for the Ecosystems of the Barents Sea as a Result of Nitrogen Oxide Emissions During the Operation of the Shtokman Field

3.1.1.1 Hazard Identification

Due to the short residence time of nitrogen oxides in the atmosphere and their active removal by dry and wet deposition on the underlying surface, the emission of these compounds when using gas for technological needs at gas industry facilities poses a threat of increased nitrogen supply to ecosystems in relative proximity to emission sources (Mingulov et al., 2005). The specifics of depositing air pollutants in the waters of the Arctic seas as a result of the exploitation of deposits in the shelf zone are associated with their accumulation most of the year (up to 8–9 months) in the surface ice and snow cover, which is characterized by movement in space relative to the source of emissions in accordance with the general direction of drifting ice. Since for the Barents Sea the main direction of currents is from west to east and north (Fig. 3.2), the zone of impact of emissions from the planned technological facilities of the Shtokman field will be "shifted" toward the coast of Severnaya Zemlya. The chemical composition of surface ice is characterized by less mineralization compared to marine waters due to the contribution of precipitation. In the spring, pollutants accumulated on the surface of the water area, mixing with marine waters, are redistributed at some depth. This mixing layer can be 10–50 m from the surface (Matishov & Dezhnyuk, 2007). In summer, the redistribution of pollutants in the upper layer of water masses occurs constantly, which must be taken into account when quantifying risks.

Environmental risks for marine ecosystems as a result of nitrogen oxide emissions associated with the combustion of gas during its production at the Shtokman field are due to the general nature of the effects of this group of pollutants, namely, their acidifying and eutrophying effect on the biogeochemical environment. Taking into account the existing indicators of salinity (35%%) and alkalinity (pH = 8–8.5) of the waters of the Barents Sea (Matishov & Dezhnyuk, 2007), the acidifying effect can be neglected, while the effects of eutrophication can pose an environmental hazard to the normal functioning of marine ecosystems.

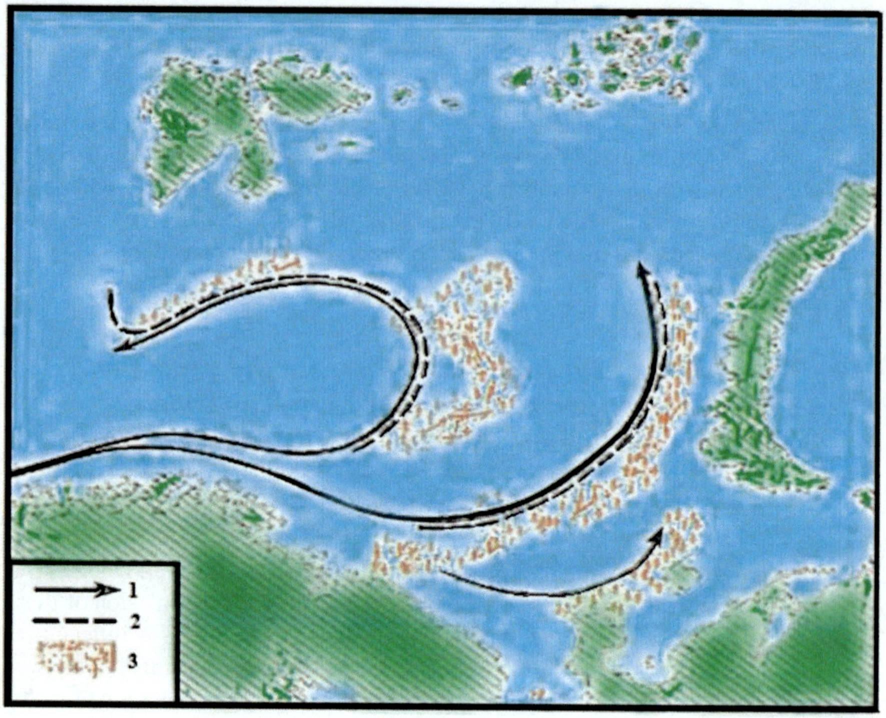

Fig. 3.2 Diagram of the directions of the main sea currents (1), polar front zones (2), and areas of increased biomass (3) in the Barents Sea (Zenkevich, 1963)

It is known from the literature that an increase in nitrogen deposition levels compared to the natural background affects all components of the nitrogen mass balance and associated macronutrients, which include carbon, oxygen, phosphorus, as well as magnesium, sodium, and calcium cations. As a rule, the natural regulation of the nitrogen cycle occurs due to an increase in the capacity of a small biological cycle because of an increase in the productivity of biotic components of ecosystems. The limiting factor for the intensification of the biological nitrogen cycle in marine and other types of aquatic ecosystems is phosphorus (Hessen et al., 1992; Moiseenko, 2003; Bashkin, 2006a, b). With a lack of phosphorus, nitrogen entering marine waters is not completely fixed by nitrogen fixators (the need for phosphorus is higher than that of biota that does not fix nitrogen), and its excess leads to an increase in the concentration of mineral nitrogen compounds in marine waters (Vitousek & Howard, 1991; Howarth, 1996; Bashkin, 2002, 2014). Other mechanisms of "stabilization" of the nitrogen cycle in marine ecosystems under conditions of increased anthropogenic supplies are accelerated denitrification at depth and an increase in the pool of nitrogen entering the bottom sediments due to the deposition of dying organic matter. The sequence of transformation processes of technogenic nitrogen compounds for marine ecosystems can be expressed by the scheme shown in Fig. 3.3.

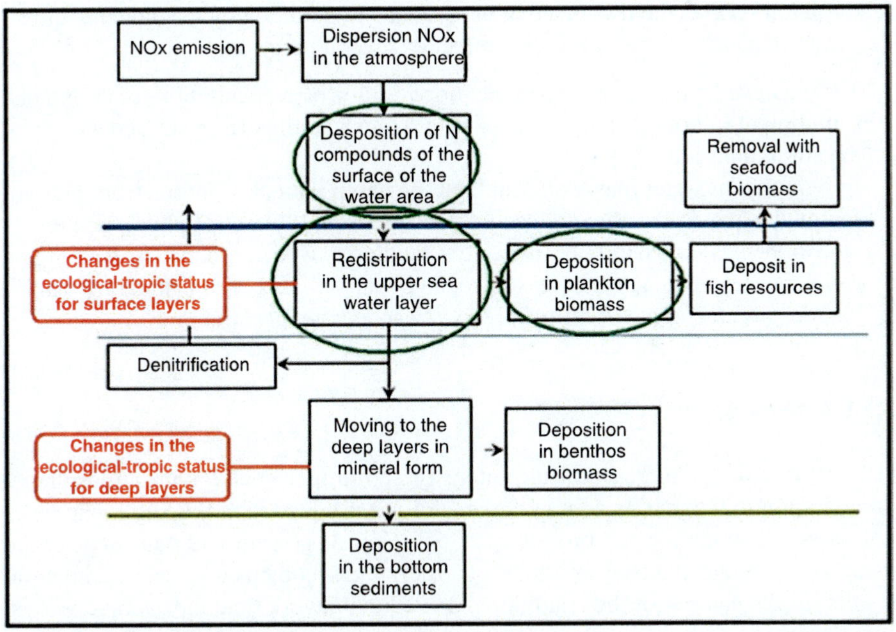

Fig. 3.3 Diagram of the "event tree" for assessing environmental risks resulting from man-made NO_x emissions from offshore natural gas production (possible effects of environmental violations are shown in red and flows included in risk calculations are shown in green)

Biological and biogeochemical indicators and processes characterizing the presence of impacts and the corresponding environmental risks are as follows:

- Changes in ecological and trophic parameters in the surface layers of marine waters
- Increased concentrations of total *N* and nitrates
- Change in the chlorophyll index
- Increase in the ratio of macronutrients (*N*/*P*, *N*/*C*, etc.)
- Changes in the biomass and productivity of phytoplankton and zooplankton in the surface water layer
- Increase of denitrification parameters in deep water layers
- Changes in ecological and trophic parameters in the bottom zone
- Deterioration of water transparency
- Reduction of free oxygen concentrations
- Changes in biomass and productivity of benthic communities

The existing classifications of marine waters according to their biogeochemical characteristics reflect the high variability of natural fluctuations of most of the parameters considered (Baalsrud, 1990; Hessen et al., 1992). In addition, according to oceanological studies (Matishov & Dezhnyuk, 2007), the Barents Sea is characterized by active vertical circulation of water masses, which can be considered as a factor that increases the resistance of ecosystems to pollution by technogenic nitrogen compounds due to the combined effect of several processes:

- Mixing of water masses and redistribution/dilution of concentrations of pollutants
- Enrichment of bottom water masses with oxygen coming from surface waters in the autumn period
- Enrichment of water masses throughout the depth with phosphorus from bottom sediments, which increases the ecological and trophic status of phytoplankton
- Due to the general salinity of the waters, leading to a decrease in dissolved nitrogen concentrations, since excess nitrogen in salty waters is consumed by plankton, despite the high *N*/*P* ratio and low phosphorus concentrations

3.1.1.2 Assessment of Exposure

Calculations of the increase in the intensity of nitrogen intake with atmospheric precipitation to the marine area in the area of gas production at the Shtokman field are based on modeling the processes of emission dispersion and NO_x deposition using data on the planned volumes of process gas combustion. The simulation results for the squares of the conditional 1 × 1 km network (and the average values for the squares of the 10 × 10 km network) show that the impact zone affects a section of the water area about 15 km in radius from the source of emissions. However, taking into account the movement of water masses, the real impact zone will be shifted in space to the northeast toward the Northern Earth. If it is necessary to obtain more accurate data on the propagation of the impact zone in space, special calculations of the movement of water masses in the considered area of the Barents Sea should be included in the model estimates.

According to the performed model calculations, the level of nitrogen deposition along the periphery of the potential impact zone after the commissioning of the Shtokman field will increase by about 3–5 kg N/ha per year compared to the background, whereas at a distance of 4–5 km from the source of emissions, atmospheric intake may be more than 100 kg N/ha per year (Fig. 3.4).

For comparison, the background nitrogen intake with atmospheric precipitation in the Arctic seas is estimated by experts at the level of 5 kg N/ha per year, and for the seas of Northern Europe, which were under conditions of more intense anthropogenic loads, they reached, on average, 20–50 kg N/ha per year (Hessen et al., 1992).

3.1.1.3 Analysis of Biogeochemical Indicators Characterizing the Dose-Effect Relationship for Marine Ecosystems

In the ecology of marine ecosystems, the most important attention is currently attached to the development of biological indices for planktonic and benthic communities. Their "response" to negative effects as known may be more informative

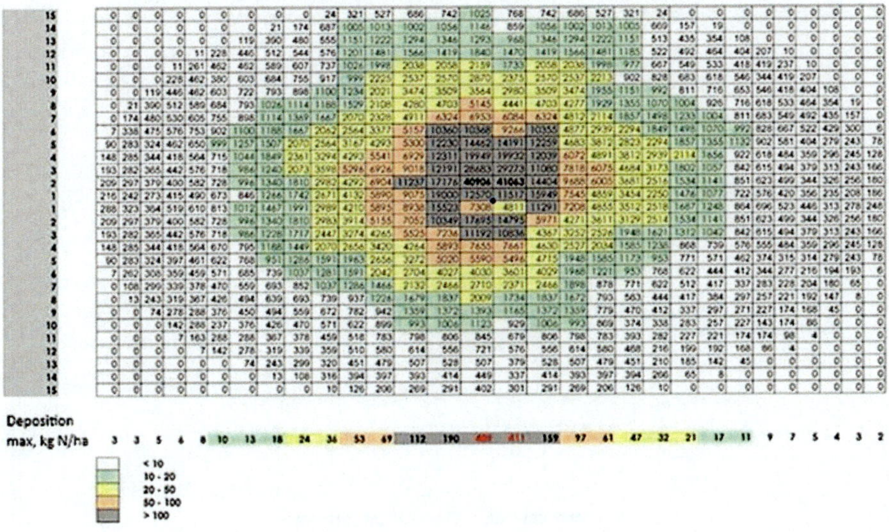

Fig. 3.4 Calculated N deposition rates of the surrounding marine ecosystems in the Shtockman GCF operation region (Priputina & Bashkin, 2017)

than data on changes in the chemical characteristics of water masses and sediments (Lerberg et al., 2000; Bashkin, 2006a, b; Sigareva, 2013; Vet et al., 2014; Moiseenko et al., 2015, 2017). As a rule, the choice is made toward macrobenthos as the main indicator criterion, which is due to a number of reasons:

- These species live near the surface of bottom sediments, where the effects of pollution and lack of oxygen are most noticeable.
- Benthos, as a rule, is sedentary and reacts to changes in the quality of the environment.
- Many benthos species are relatively long-lived species, which allows us to assess violations over time.

There are many species in the benthos with different "life strategies," which will allow them to be classified into different functional groups:

- Some types of benthos are of commercial importance.
- These species are exposed to pollutant flows at the interface between water and sediments due to bioturbation and suspension.

As a rule, assessments of the "disturbance" of marine ecosystems using biological indicators are empirical, since they are based on monitoring observations and have the character of point estimates or comparative analysis of environmental indices (Pinto et al., 2009). However, the impact of anthropogenic air pollution on the state of marine biota occurs as a result of a sequential "chain" of quantitative changes in the marine environment affecting both biological and hydrochemical indicators (Fig. 3.5).

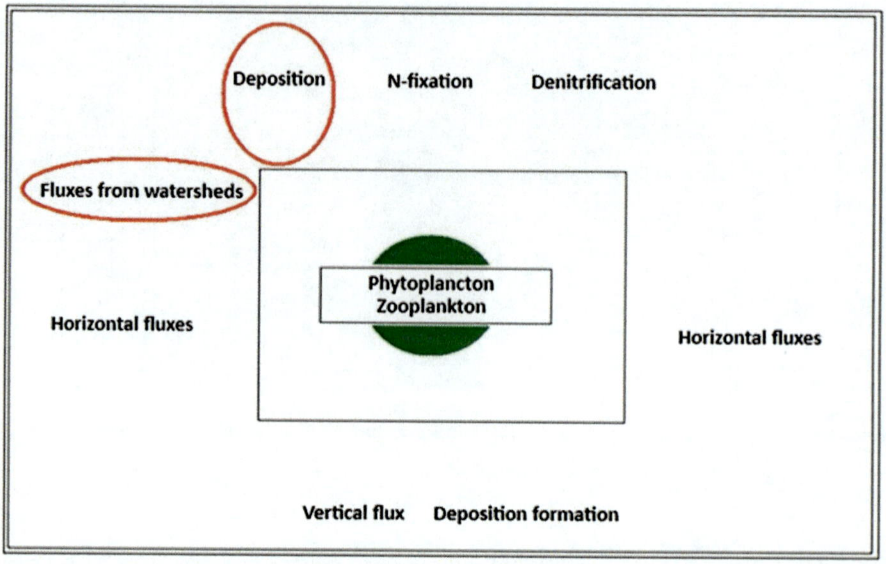

Fig. 3.5 Risks to marine ecosystems from impacts of atmospheric NO_x emissions from offshore gas production

Environmental risks are associated with the effects of nitrogen oxide emissions:

- Eutrophication
- Decrease in O_2 concentrations
- *N*:*P*:*C* imbalance

3.1.1.4 Risks of Violations of the Ecological and Trophic Status of Waters in the Benthic Development Zone

These risks in relation to the impacts associated with an increase in nitrogen supplies are determined primarily by the effects of "oxygen deficiency," which is determined by an increase in the consumption of free oxygen for the oxidation of organic matter resulting from the deposition of dead plankton biomass, as well as the cost of oxygen for nitrification. According to available data, 222 g of O_2 is needed for the oxidation of 1 kg of raw zooplankton mass. Correspondingly (Zenkevich, 1963), the amount of zooplankton in the open part of the Barents Sea is 200–2000 mg/m^3, which for a 50-meter layer corresponds to 40–400 kg/ha. According to our estimates, a possible increase in zooplankton reserves due to an increase in phytoplankton production at the calculated levels of nitrogen deposition can reach 10%, increasing by an appropriate amount the oxygen consumption required for the oxidation of organic matter after the biota dies off (Fig. 3.6).

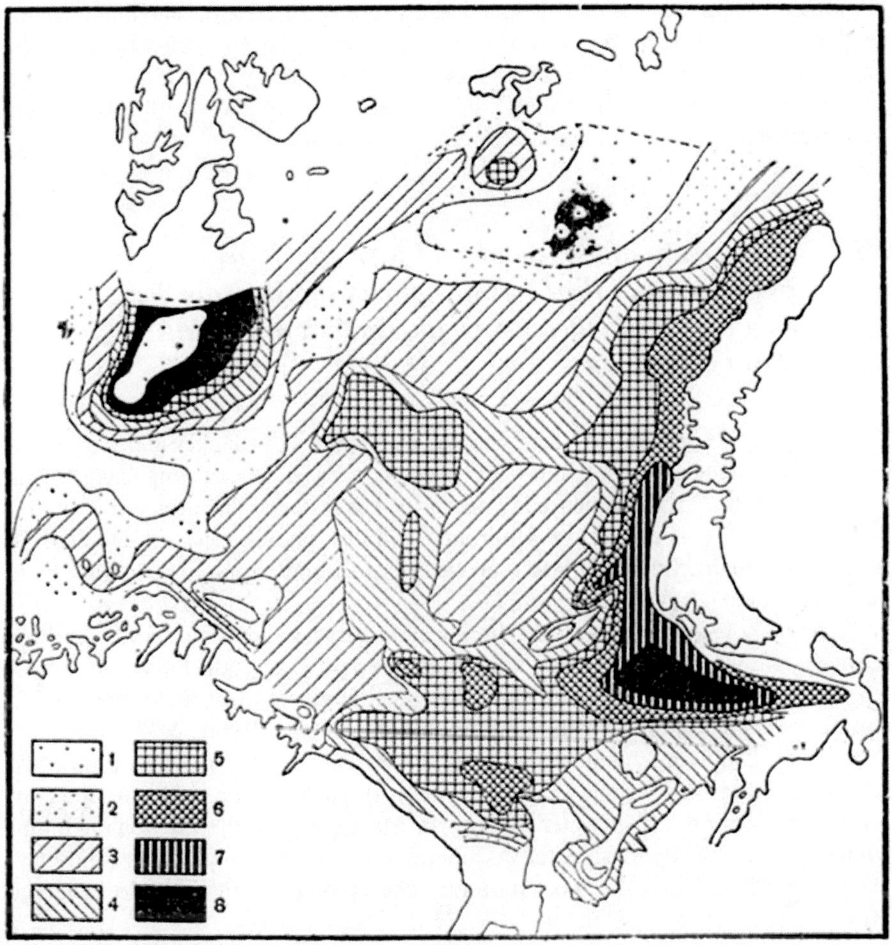

Fig. 3.6 Effect of nitrogen inputs on phytomass production: benthic biomass distribution in the Barents Sea (g/m^2): 1 – < 10; 2 – 10-25; 3 – 25-50; 4 – 50-100; 5 – 100-200; 6 – 200-300; 7 – 300-500; 8 – > 500 (Zenkevich, 1963)

This oxygen will be consumed by the biota during biological denitrification. Up to 138 g of oxygen is required to create 1 mol of N_2 with this type of denitrification, the parameters of which in marine ecosystems are estimated by experts at 0.1–4 mg N/m^2 per day (or from 0.3–0.4 to 14–15 kg N/ha per year). A multiple increase in nitrogen deposition levels compared to the background in the zone of the greatest impact of the Shtokman deposit, leading to an increase in dissolved nitrogen concentrations in marine waters, will contribute to the intensification of denitrification processes, and, accordingly, an increase in environmental risk for the species composition and the state of the benthic community.

However, as in the case of phytoplankton, spatially the zone of potential risks for benthic communities in the first years of operation of the Shtokman field will be limited to a section of the water area with a radius of 15 (±25 km, taking into account the influence of sea currents). In the long term, cumulative effects of disturbances are not excluded for benthic communities.

3.1.2 Assessment of Geoecological Risks for Marine Ecosystems of the Coastal Zone of the Barents Sea as a Result of Nitrogen Oxide Emissions During Plant Operation

To characterize geoecological risks, coupled calculations of changes in these parameters at specified exposure values are necessary, based on existing data on the parameters of the normal functioning of marine ecosystems. The literature provides data on optimal concentrations of nitrogen, nitrates, phosphorus, and chlorophyll for different hydrochemical classes of waters and trophic groups of marine biota (Tables 3.1 and 3.2).

Some authors suggest using the atomic ratio between the main biofils (carbon, nitrogen, and phosphorus) under "normal conditions" as an indicator when assessing the risks of an imbalance of nutrients. A different ratio of these elements in marine waters will indicate saturation or limitation of N or P (Fig. 3.7).

According to hydrochemical studies in the Barents Sea (Zenkevich, 1963), two relatively contrasting periods of the year are distinguished with respect to the concentrations of nutrients in water masses. In March, due to the previous autumn-winter mixing of water masses, the distribution of nutrient compounds (nitrates and phosphates) in them is the most uniform. The amount of nitrates is on average

Table 3.1 Concentrations of nutrients, oxygen, and chlorophyll (mkg/l) for different types of marine waters (according to Baalsrud 1990; Hessen et al., 1992; Bashkin & Priputina, 2010)

	I class	II class	III class	IV class
	Mean for the summer period			
N_{tot}	<250	250–330	330–500	>500
$N\text{-}NO_3$	<12	12–23	23–65	>65
P_{tot}	<12	12–16	16–29	>29
Chlorophyll	<1.9	1.9–3.4	3.4–7.3	>7.3
	Mean for the winter period			
N_{tot}	<295	295–380	380–560	>560
$N\text{-}NO_3$	<90	90–125	125–225	>225
P_{tot}	<21	21–25	25–42	>42
	Mean for the year			
O_2 (mean)	>5.3	5.3–3.8	1.0–0.1	<0.1
O_2 (min)	>3.2	3.2–1.0	0.3–0.1	<0.1

150–250 mg/m^3 (or 35–55 mkg N/l) for the Barents Sea, which corresponds to the conditions characteristic of eutrophic communities. Phosphate concentrations are estimated at 40–60 mg/m^3. In the spring-summer period, due to the active development of phytoplankton in the upper layers of waters, the amount of nitrates drops to almost 0, amounting to 200 mg/m^3 in the bottom layer. Phosphorus deficiency is also characteristic of the upper layers in the summer, while at depth, its concentrations at this time are estimated at 60 mg/m^3. In the autumn period, when the bottom masses move from deep layers to the surface, the composition of nutrient salts is restored throughout the vertical profile of water masses.

The estimated impact zone of nitrogen oxide emissions accompanying gas production at the Shtokman field refers to a section of the Barents Sea with a relatively low benthic productivity of 25–50 g/m^2 (up to 100 g/m^2). However, the existing system of currents determines the transit of water masses with pollutants deposited in them to the east direction forwarded to the coast of the Northern Land, where the productivity of marine ecosystems is higher – 200–500 g/m^2. According to existing concepts based on experimental data (UBA, 2004), more productive ecosystems are characterized by greater resistance to the effects of eutrophication associated with an increase in anthropogenic nitrogen loads.

Table 3.2 Nitrogen concentrations (mkg/l) for different trophic groups of marine biota (Wulff, 1990)

	Oligotrophs	Mesotrophs	Eutrophes	Hypereutrophes
N_{tot}	<250	250–350	350–450	>450
N-NO_3	<10	10–30	30–50	>50

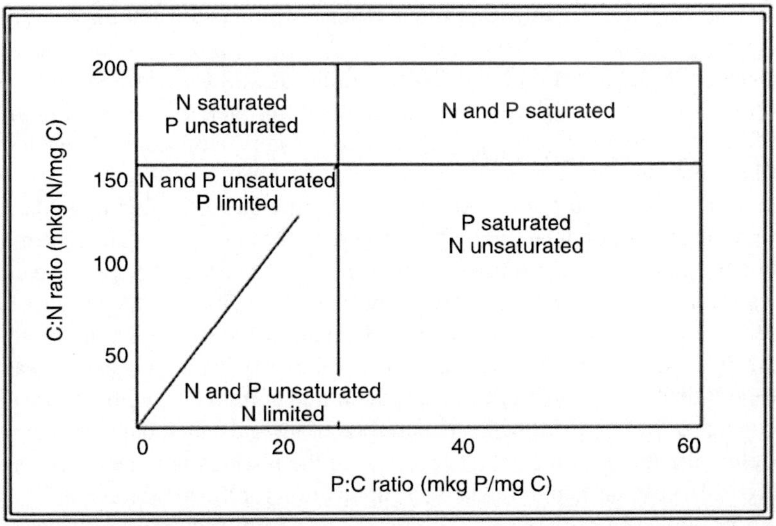

Fig. 3.7 The nitrogen/phosphorus ratio and the corresponding ecological and trophic status of various seaweed groups (Vadstein et al., 1989; Howarth, 1996; Bashkin, 2006a, b)

3.1.2.1 Risk Characterization

Quantitative assessments of environmental risks for the NO_x emission zone during the operation of the Shtokman field were carried out taking into account the preservation of optimal ecological and trophic conditions for phytoplankton, which is the initial link of the biological nitrogen cycle in marine ecosystems. Qualitative risk assessments were carried out for benthos communities.

Risk assessments of changes in the trophic status of marine waters in the phyto- and zooplankton development zone were carried out in accordance with the following scenario. It was believed that in winter (8 months a year), nitrogen is deposited in the surface ice and snow cover, the thickness of which can be 2–5 m. The calculated concentrations of nitrogen entering the water area corresponding to the model levels of its precipitation are presented in Table 3.3. Nitrogen accumulated on the surface of the water area during the winter period determines the short-term "peak" of eutrophication of marine ecosystems, since the melting of ice coincides with the mass spring development of phytoplankton (Zenkevich, 1963). Since at this time marine waters are not limited in phosphorus, all nitrogen can be included in the biological cycle as part of phytoplankton, and the saturation of surface waters with oxygen (more than 100%) is a "guarantee" of maintaining a favorable oxygen regime in them.

During the summer period (4 months a year), due to the intake of nitrogen with atmospheric precipitation to the surface of the water area and subsequent mixing of the upper layer of water, a gradual increase in nitrogen concentrations in this part of the sea column occurs (mixing depth is 10–50 m). The calculations performed show that the total level of nitrogen concentrations during the growing season (excluding its consumption by the biota), depending on the amount of precipitation and the thickness of the water column, can range from 3–16 to 260–1300 micrograms N/l (Table 3.3). It has already been mentioned that according to field observations (Zenkevich, 1963), there are practically no nitrates in surface marine waters during this period. Thus, the calculated concentrations of nitrogen (represented mainly by nitrate forms) will determine the trophic status of waters in the zone of development of plankton communities.

Taking into account the selected criterion and the data obtained in Table 3.4, at the level of precipitation up to 10 kg N/ha per year, the environmental risks of eutrophication for plankton communities in this section of the Barents Sea can be considered negligible (green). The level of precipitation of 10–50 kg N/ha per year can be considered as a low level of risk (orange). With an intake of more than 50 kg N/ha per year, it is possible to change the trophic status of waters and the appearance/increase of the share of hypereutrophs in the biotic community with a reduction of oligotrophic species. According to the analysis of the presented data and the results of modeling the levels of nitrogen deposition at the planned volume of gas production (Fig. 3.4), environmental risks of eutrophication for the plankton zone are most likely to be observed within a 10-kilometer zone from the emission source (red, unfavorable scenario).

Table 3.3 Increase in nitrogen content in marine waters and ice and snow cover as a result of an increase in nitrogen deposition levels during the operation of the Shtokman field

Precipitation rate (kg N/ha per year)	Increase in the concentration of *N* in the sea water layer (mkg/l per day)		The total increase in the concentration of *N* in the sea water layer during the summer season (mkg/l)		Accumulation of *N* in the ice and snow cover during the winter season (mg/m^3)	
	10 m	50 m	10 m	50 m	in layer 2 m	in layer 5 m
5	0.14	0.03	16	3	168	67
7.5	0.21	0.04	25	5	252	101
10	0.27	0.05	33	7	336	134
15	0.41	0.08	49	10	503	201
20	0.55	0.11	66	13	671	268
30	0.82	0.16	99	20	1007	403
50	1.37	0.27	164	33	1678	671
100	2.74	0.55	329	66	3356	1342
150	4.11	0.82	493	99	5034	2014
200	5.48	1.10	658	132	6712	2685
400	10.96	2.19	1315	263	13425	5370

However, the calculations made do not take into account the movement of water masses (and the natural dilution of nitrogen concentrations in this case), as well as the active fixation of available nitrogen contained in marine waters in phytoplankton biomass, which reduces its content in water. According to the available expert estimates for the North Sea, the daily total nitrogen intake with precipitation at the level of 0.2 mkg N/l per day can increase phytoplankton biomass by an amount corresponding to 1.2 mg C/m^3 (Hessen et al., 1992). Assuming that all nitrogen received from atmospheric precipitation during the summer period can be absorbed by biota, it is possible to calculate the increase in phytoplankton mass reserves at different levels of nitrogen deposition (Table 3.4).

Table 3.4 An increase in phytoplankton biomass reserves as a result of an increase in nitrogen deposition levels during the operation of the Shtokman field

Precipitation rate (kg N/ha per year)	Additional supply *N* for 50 mm layer (mkg/l)	[a]Increase of C reserves in phytoplankton biomass for a 50 m layer (mg C/m^3)	Additional intake *N* for 10 m layer (mkg/l)	Increase of C reserves in phytoplankton biomass for the 10 m layer (mg C/m^3)
5.0	3	0.4	16	2
7.5	5	0.6	25	3
10	7	0.8	33	4
15	10	1.2	49	6
20	13	1.6	66	8
30	20	2.4	99	12
50	33	4	164	20
100	66	8	329	39
150	99	12	493	59
200	132	16	658	79
400	263	32	1315	158

[a]In the calculations performed, the ratio between the reserves and the productivity of phytoplankton biomass, characteristic of the Barents Sea, was used, equal to 50 (Zenkevich, 1963)

Taking into account the elemental composition of phytoplankton, in which nitrogen accounts for about 10% (Zenkevich, 1963), the reserves of nitrogen fixed in phytoplankton biomass in terms of area will increase on average over the summer period for the zone of maximum impact by 1–1.5 kg N/ha.

The presented calculations of the increase in productivity and phytoplankton stocks because of the increase in atmospheric nitrogen supply do not take into account the limitation of nitrogen involvement in the biological cycle characteristic of the summer period by phosphorus. The sources of phosphorus for phytoplankton are the deep waters rising to the surface of the water area in the autumn-winter period and the intake with atmospheric precipitation. The literature presents contradictory data on the atmospheric supply of phosphorus. Thus, in a number of publications, the *N*/*P* ratio in atmospheric precipitation of background territories is estimated at 150:1, which, at a nitrogen deposition rate of 5 kg N/ha per year (background intake to the Arctic seas) corresponds to 3.5 kg P/km^2 per year (Vet et al., 2014). For the marine areas of the Nordic countries, the value of 10 kg P/km^2 per year is indicated (Hessen et al., 1992), which, taking into account the contribution of marine aerosols to the total component of precipitation, seems more reliable. Using the "normal" for phytoplankton atomic ratio *C*:*N*:*P* = 106:7:1 (Redfield, 1958), it is possible to calculate the amount of nitrogen that can be additionally absorbed by phytoplankton during the summer season due to atmospheric phosphorus supply.

Also, using the "normal" atomic ratio *C*:*N*:*P* = 106:7:1 for phytoplankton, it is possible to calculate the amount of oxygen required to create its biomass, equal to 8.6 mol (Redfield, 1958):

$$(CH_2O)_{106}(NH_3)_{16}(H_3PO_4) + 138\,O_2$$

$$106\,CO_2 + 16\,HNO_3 + H_3PO_4 + 122\,H_2O$$

Assuming that all nitrogen entering ecosystems will be assimilated by biota, it is possible to calculate the amount of oxygen necessary for the creation of biomass, the lack of which leads to the risks of death of fish resources as a result of oxygen deficiency for their respiration:

$$orgN_{aerob} = orgN_{used} - orgN_{immob} - orgN_{export} - orgN_{anaerob}$$

$$O_{2\,input.} < \left(orgN_{used} - orgN_{immob} - orgN_{export} - orgN_{anaerob}\right) * 8.6$$

It is less than 1 kg N/ha, which determines the risk of *N*/*P* imbalance for phytoplankton at the end of the growing season, when the "deep" phosphorus reserve in the surface layer of marine waters will be used up for almost the entire impact zone of the Shtokman deposit.

The risks of violations of the ecological and trophic status of waters in the benthic development zone in relation to the effects associated with an increase in nitrogen supplies are determined primarily by the effects of "oxygen deficiency," which is determined by an increase in the consumption of free oxygen for the oxidation of organic matter resulting from the deposition of dead plankton biomass, as well as oxygen costs for nitrification. According to available data, 222 g of O_2 is needed for the oxidation of 1 kg of raw zooplankton mass. According to available data (Zenkevich, 1963), the amount of zooplankton in the open part of the Barents Sea is 200–2000 mg/m^3, which for a 50-meter layer corresponds to 40–400 kg/ha. According to our estimates, a possible increase in zooplankton reserves due to an increase in phytoplankton production at the calculated levels of nitrogen deposition can reach 10%, increasing by an appropriate amount the

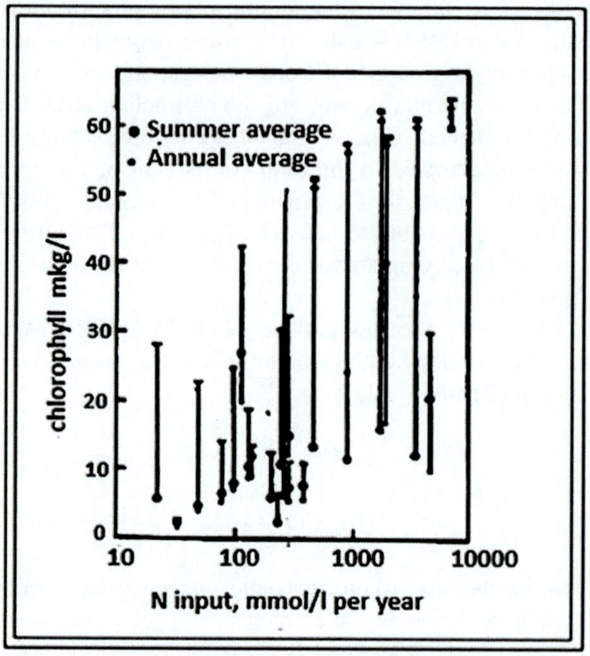

Fig. 3.8 Effect of nitrogen inputs on phytomass production: the ratio between the concentration of chlorophyll and the annual nitrogen intake according to monitoring and experimental studies (Nixon & Pilson, 1983)

oxygen consumption necessary for the oxidation of organic matter after the death of biota.

This oxygen will be consumed by the biota during biological denitrification. Up to 138 g of oxygen is required to create 1 mol of N_2 with this type of denitrification, the parameters of which in marine ecosystems are estimated by experts at 0.1–4 mg N/m^2 per day (or from 0.3–0.4 to 14–15 kg N/ha per year). A multiple increase in nitrogen deposition levels compared to the background in the zone of the greatest impact of the Shtokman deposit, leading to an increase in dissolved nitrogen concentrations in marine waters, will contribute to the intensification of denitrification processes, and, accordingly, an increase in environmental risk for the species composition and the state of the benthic community (Fig. 3.8).

However, as in the case of phytoplankton, spatially the zone of potential risks for benthic communities in the first years of operation of the Shtokman field will be limited to a section of the water area with a radius of 15 (±25 km, taking into account the influence of sea currents). However, there are possible short-term "nutritional stress" and long-term (systematic) changes in the ratio of food elements (Fig. 3.9). Accordingly, in the long term, cumulative effects of disturbances are not excluded for benthic communities.

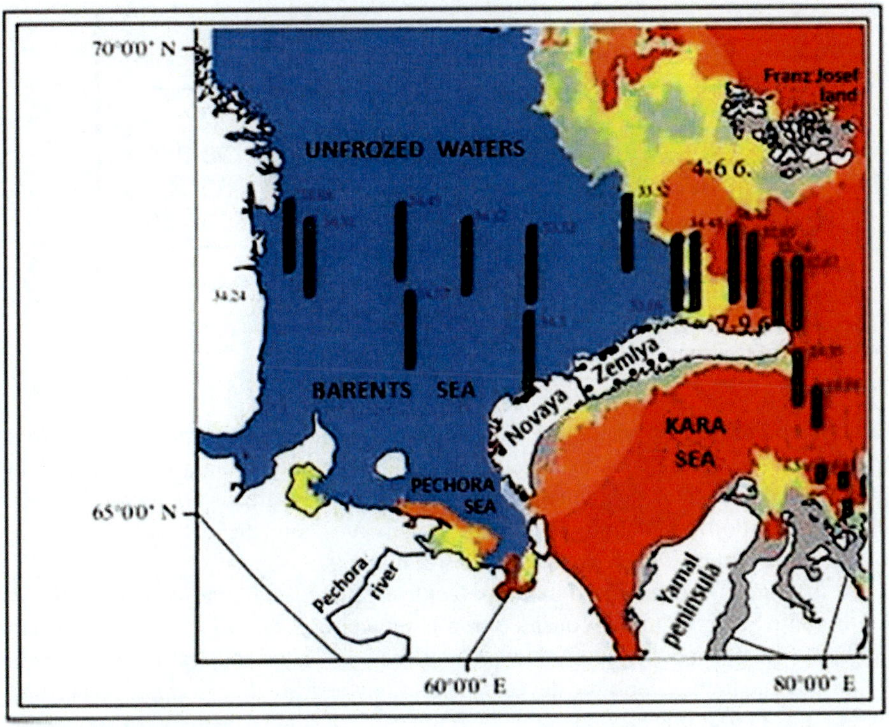

Fig. 3.9 Nutrient imbalance indicators: the resistance of marine ecosystems with higher salinity to excess nitrogen is higher due to a more active supply of *P* from the depths, which leads to the preservation of the *N*/*P* balance necessary for active nitrogen assimilation by plankton

3.1.3 *Assessment of Geoecological Risks for Marine Ecosystems of the Coastal Zone of the Barents Sea as a Result of the Emission of Nitrogen Oxides During the Operation of the LNG Plant*

3.1.3.1 Hazard Identification

The environmental hazard factors for marine ecosystems of the littoral zone due to the emission of nitrogen oxides during the operation of the planned LNG plant are generally similar to those discussed above. The main differences are related to the peculiarities of the functioning of the coastal areas of the seas: their shallow depth and tidal currents (water level fluctuations for the considered zone are 4 m). Consequently, the effects of "dilution" of anthropogenic loads of atmospheric pollutants in this part of the marine area are less pronounced. At the same time, the littoral zone belongs to the most productive ecosystems of the seas, which, with respect to the effects of eutrophication, increases the stability of these ecosystems compared to the ecosystems of the main water area with equal levels of loads of

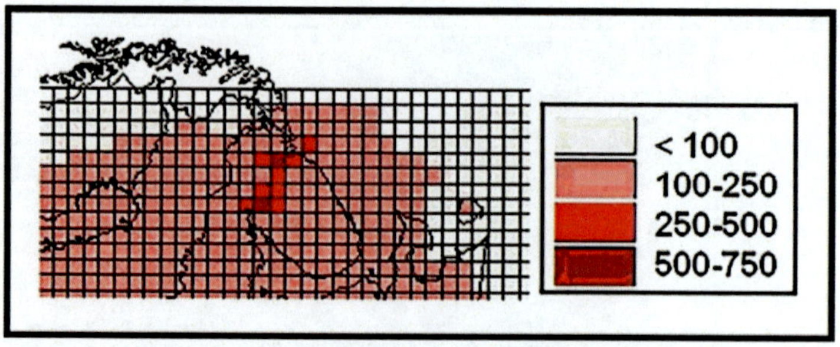

Fig. 3.10 Results of modeling of sulfur deposition for the Kola region (g-eq./ha per year; calculated square of modeling 50 × 50 km)

technogenic nitrogen compounds. This is the richest zone of littoral fauna. Littoral biomass can be up to 24–28 kg/m^2. The ratio of annual phytoplankton production in littoral zone to its average annual biomass in the open sea area is 50, for zoobenthos 0.2–0.25. Bottom sediments are enriched with nitrogen, the *C/N* ratio is 5.5–8, which exceeds the level characteristic of soil humus.

At the same time, an additional risk factor for the coastal marine ecosystems of the Barents Sea area under consideration is the contribution of coastal runoff and atmospheric transport of pollutants, which has an increased level on the Kola Peninsula due to the economic development of the territory. For example, according to the data of modeling carried out in the framework of studies on the UN Convention on the Long-Range Transboundary Air Pollutants (CCE Report, 2005; Dentener et al., 2006; Garmo et al., 2014; Moiseenko et al., 2015, 2016), the existing level of acid precipitation for the territory of the Kola Peninsula is estimated at 200–250 g-eq./ha per year (Fig. 3.10).

For the site under consideration, located in the zone of atmospheric transport of pollutants from the enterprises of Monchegorsk and Murmansk, it reaches, according to minimum estimates, 500 g-eq./ha per year, which corresponds to the intake of 6–7 kg S/ha per year. The existing level of nitrogen deposition for the Kola Peninsula according to the same data (CCE Report, 2005) is estimated at 3–5 kg N/ha per year.

3.1.3.2 Exposure Assessment

The model estimates of nitrogen deposition levels include three scenarios of the design capacity of the planned LNG plant (3 million tons per year, 7.5–8 million tons per year, and 15 million tons per year). Figure 3.11 shows the layout in the space of the exposure zone for the minimum and maximum NO_x emission scenarios. The cell of the simulation square corresponds to an area of 1 × 1 km^2. As can be seen from the diagram, the main impact zone falls on the coastal part of the Kola Peninsula, extending its peripheral part to the sea area for about 5 km. Consequently, for littoral ecosystems, the level of additional nitrogen deposition should be

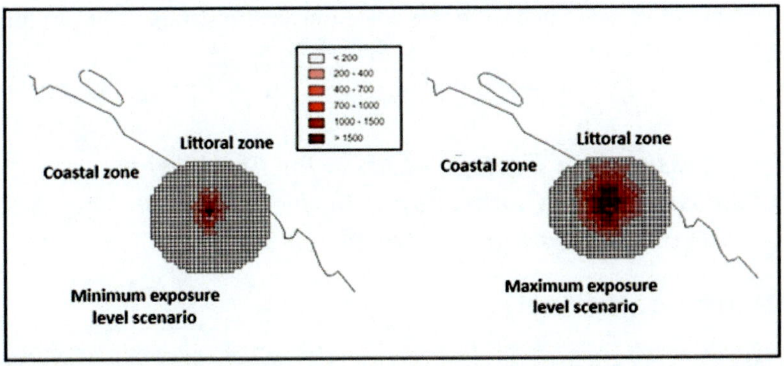

Fig. 3.11 Layout of the impact zone of the LNG plant (the level of nitrogen deposition for the simulation cell is 1 × 1 km-g-eq/ha per year)

relatively small – in the range of 3–10 kg N/ha per year, depending on the intensity of the scenario (Fig. 3.11).

3.1.3.3 Analysis of the Dose-Effect Relationship for Littoral Ecosystems and Characterization of Potential Environmental Risks

The data characterizing the normal functioning and preservation of the biotic structure of marine ecosystems and their resistance to anthropogenic disturbance of the established nitrogen cycle are applicable to the littoral zone. The necessary adjustments should be made to the specifics of the migration of water masses in the coastal zone, the lower depths in this part of the sea area, and the possible contribution of coastal runoff to the balance of nutrients.

According to data (Zenkevich, 1963), macrophyte biomass reserves in the sublittoral (represented mainly by various forms of kelp at the surface and at depth by red algae) average 10 kg/km^2 (the variations are from 1–3 to 27 kg/km^2). Taking into account the calculated levels of additional nitrogen intake and the data in Table 3.4, the increase in biomass for these ecosystems will be less than 1%, which should not affect a significant increase in the "costs" of oxygen for the mineralization of organic matter.

In the absence of other sources of atmospheric pollution, the geoecological risks associated with the emission of nitrogen oxides into the atmosphere and increased levels of nitrogen deposition on the coastal part of the Barents Sea, with respect to the effects of acidification and eutrophication, could be considered as minimal or permissible (depending on the exposure scenario). However, with the existing high level and diversity of anthropogenic impacts in this region, for accurate assessments, special field studies are needed to assess the degree of disturbance of these ecosystems and possible features of biogeochemical nitrogen flows in the ebb-tide zone. It is possible that under these conditions the role of denitrification processes in the regulation of the nitrogen cycle will increase, because at the time of low tide, bottom sediments enriched with organic matter and mineral forms of nitrogen in exposed areas of the sea can be a source of significant emissions of nitrous forms of nitrogen.

3.1.4 Assessment of Ecological Risks for Freshwater Ecosystems in Connection with the Operation of the LNG Plant in the Far North

3.1.4.1 Hazard Identification

In the absence of industrial and agricultural emissions of nitrogen compounds, concentrations of nitrates and ammonium forms in lakes and swamps of the tundra zone, to which most of the impact zone belongs, have low values (Moiseenko, 2003, 2005; Moiseenko et al., 2016). With an increase in the atmospheric supply of nitrogen to catchments, first, the effects of acidification of freshwater reservoirs arise. Moreover, the increased intake of ammonium compounds may also be the cause of acidification because of the development of nitrification processes leading to the release of additional protons.

Additional factors of increasing the acidity of reservoirs are the processes of accelerated dying and mineralization of organic litter in the catchment area during technogenic transformation and violations of terrestrial vegetation cover, leading to increased concentrations of organic acids in the runoff of soil and groundwater (Bashkin et al. 2002a; Semenov, 2002; Bashkin, 2006a, b; Dongwei et al., 2022). The specifics of the weather and climatic conditions of this region determine the deposition of atmospheric pollutants entering the catchment areas in the snow cover for a significant part of the year (Trubitsina, 2016a, b). The main melting of snow in spring and the redistribution of meltwater from catchments to local runoff zones (freshwater lakes and swamps) occur in frozen soils, which determines the practical absence of neutralization and transformation of pollutants accumulated in the snow cover. In addition, the granite-gneiss composition of soil-forming rocks within the considered part of the Kola Peninsula belongs to the most acidic sediments with low buffering capacity in terms of neutralizing the acid component of atmospheric precipitation (Belov & Baranovskaya, 1969; UBA, 2004).

Thus, the greatest environmental risks for freshwater ecosystems in the zone of impact of the LNG plant may be associated with the effects of acid disturbances.

3.1.4.2 Exposure Assessment

As shown in Fig. 3.11, it can be seen that near the source of emissions (up to 5 km), the level of additional nitrogen intake with atmospheric precipitation for the minimum scenario can be 10–20 kg N/ha per year with the level of precipitation at the epicenter being 40–50 kg N/ha per year and above. When the second stage of the LNG plant is put into operation with an estimated production capacity of 15 million tons per year (maximum scenario), the level of receipt of technogenic nitrogen compounds can be as follows: in the immediate vicinity of the plant within a radius of up to 1 km – more than 100 kg N/ha per year; within a radius of 3 km – more than 50 kg N/ ha per year; the impact zone with annual depositions more then 10 kg/ha will extend at a distance of up to 7–10 km from the emission source.

The dose-effect relationship for freshwater ecosystems can be estimated based on the calculation of critical loads (CL), similar to terrestrial ecosystems (UBA, 1996, 2004). CL is calculated with respect to the effects of acidification and eutrophication.

In the framework of the diatom model, the critical load of acidity is calculated by the following formula:

$$\mathrm{CL}(\mathrm{Ac}) = [\mathrm{Ca}]_0 / 89$$

where, CL (Ac) is the permissible level of water acidity reduction (keq./ha per year) and [Ca] is the Ca content in natural waters (mg-eq./l).

Another calculation method is based on the first-order acidity balance model:

$$\mathrm{CL}(S)_{\mathrm{max}} = L_{\mathrm{crit}} / (1 - p_5)$$

where, $\boldsymbol{p_s}$ is the retention coefficient of the element in the catchment area.

Typically, the acidity model is used for regions with upland peat bogs (Hall et al., 2003). The indicator criterion for this and a number of other models is the value of acid neutralizing capacity (ANC) of waters or $\mathrm{ANC_{crit}}$. Depending on the sensitivity of aquatic biota to acidification, different values are used for $\mathrm{ANC_{crit}}$: 0, 20, or 50 mg-eq./l. As a rule, for freshwater reservoirs of the Kola Peninsula, the value of $\mathrm{ANC_{crit}}$ is below 50 mg-eq./l or has negative values (Vandysh, 2002; Lozovik, 2006; Moiseenko et al., 2015, 2016).

In addition, the CL of acidity as a function of the main acidifying agent – sulfate ion – can be calculated based on data of the concentration of basic cations (BC) in natural waters.

$$[\mathrm{SO_4}]_o = a + b^* [\mathrm{BC}]_t$$

According to Norwegian data, the following coefficient values can be used to calculate the concentration of the anthropogenic component of the sulfate flow: $a = 3$, $b = 0.17$ (UBA, 2004).

Taking into account the magnitude of the hydrological runoff from the catchment area and the concentration of the main cations in the runoff, the acidity CL can be calculated as follows:

$$\mathrm{CL} = Q^* \left([\mathrm{BC}]_0 \; / \; (1 + k^* Q) \right.$$

where $k = 0.1$ or 0.25

With regard to the effects of eutrophication for lakes and small watercourses, the authors (Hall et al., 2003) propose to calculate the critical load of minimum nitrogen based on the following formula:

$$\mathrm{CL}(N)_{\mathrm{min}} = f^{*} N_{\mathrm{up}} + (1-r)^{*}(N_{\mathrm{i}} + N_{\mathrm{de}})$$

where, N_{up}, N_i, and N_{de} are, accordingly, the parameters of nitrogen fixation in forest biomass production in the catchment area, immobilization in the soil, and denitrification; f and r are the share of forest ecosystems and the total share of the catchment area. The value of N_{im} according to expert estimates is 11 g-eq./ha per year for standing lakes and 71 g-eq./ha per year for small watercourses. The same indicators are used for N_{de}.

Critical loads of nutrient nitrogen (CL $(N)_{nut}$) for lakes and watercourses, as a rule, are not calculated, since these ecosystems are limited by phosphorus, not nitrogen. For wetlands, empirical values of CL of nutrient nitrogen are usually used, estimated for the UK, for example, in the range of 5–12 kg N/ha per year (the average value is 10 kg N/ha per year). Norwegian experts estimate the CL of nutrient nitrogen for watered swamps at 5 kg N/ha per year.

Accordingly, the maximum permissible nitrogen load is the sum of CL of acidity and CL of minimum nitrogen:

$$\mathrm{CL}(N)_{\mathrm{max}} = \mathrm{CL}(S)_{\mathrm{max}} + \mathrm{CL}(N)_{\mathrm{min}}$$

3.1.4.3 Risk Characterization

There are no large water bodies (with an area of more than 1 km^2) within the territory corresponding to the main impact zone of nitrogen oxide emissions by the planned LNG plant. To assess the excess of permissible parameters of the intake of acid-forming compounds with atmospheric precipitation for freshwater ecosystems of the Kola Peninsula, data were used (Moiseenko, 2003), according to which the values of CL acidity for lakes should not exceed 200 g-eq./ha per year. As follows from the data shown in Fig. 3.12, this level is currently exceeded, which indicates the "exhaustion" of the natural potential of ecosystem buffering to the acid component of atmospheric precipitation.

The ecosystems of coastal areas are characterized by greater resistance to acidic acids in precipitation than the interior of the Kola Peninsula due to the increased content of sea salt cations in atmospheric precipitation.

Taking into account the previously calculated values of the permissible nitrogen intake into the terrestrial ecosystems of the tundra zone of the Kola Peninsula, the CL values of eutrophying nitrogen compounds for the territory under consideration are about 3 kg N/ha per year, which corresponds to the already existing level of precipitation. Thus, for most of the territory corresponding to the calculated exposure zone, it is possible to predict the risks of eutrophication of terrestrial

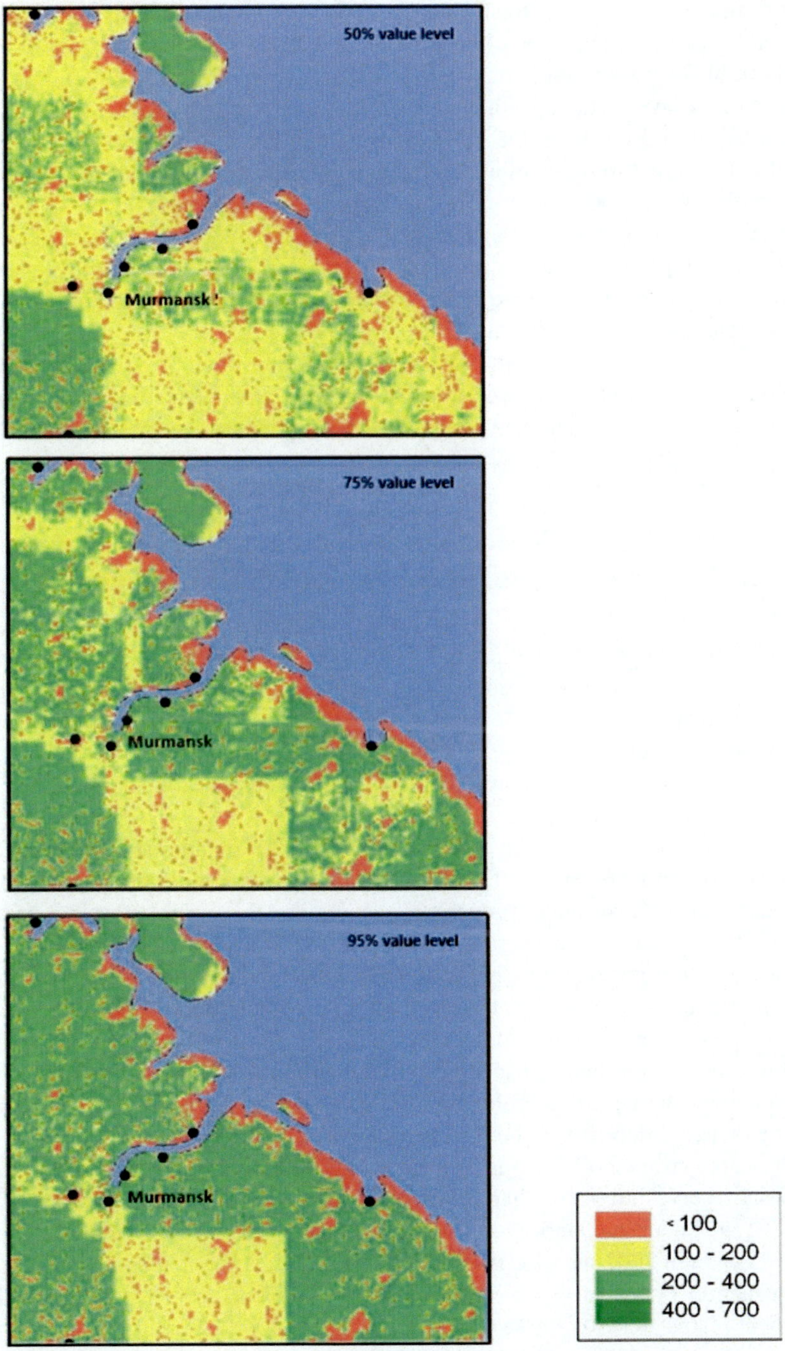

Fig. 3.12 Assessment of the stability of the ecosystems of the Kola Peninsula coast to technogenic NO_x emissions: acidification effects, g-eq/ha per year. Black dots represent different cities

ecosystems and associated freshwater reservoirs after the commissioning of the planned LNG plant. The area of the zone of increased geoecological risks of eutrophication of freshwater and terrestrial ecosystems (twofold excess of CL) will correspond to the sites removed at a distance of 5–10 km from the source of emissions. In an area up to 5 km from the LNG plant, a high level of these risks can be predicted with respect to the combined effects of acidification and eutrophication due to multiple (400 - 1000 % or more) exceeding the permissible levels of impacts during the launch of the first stage of the LNG plant (capacity – 7.5–8 million tons of LNG). Increasing the plant's productivity to 15 million tons (second stage) will "expand" the area of the territory with a high level of environmental risks to a 10-kilometer zone around the emission source (Fig. 3.13).

Thus, calculations of the values of CL of nutrient nitrogen (g-eq./ha per year) for the Kola Peninsula coast showed a reduced potential of terrestrial ecosystems in terms of assimilation of technogenic nitrogen compounds.

3.2 Assessment of Geoecological Risks in the Impact Zones of the Arctic Coastal Ecosystems of the Taz Peninsula

Currently, the main gas production in Russia takes place in the coastal regions of the Arctic. In this regard, two regions will be considered: the Taz Peninsula and the Yamal Peninsula.

GAZPROM Dobycha Yamburg, a subsidiary of Gazprom PJSC, is the largest on the Taz Peninsula, and its gas production since 1986 has already amounted to 6 trillion m^3 in mid-June 2021.

Gazprom Dobycha Yamburg LLC holds licenses for exploration and production of hydrocarbons from unique fields: Yamburgsky and Zapolyarny. A number of other deposits are also being prepared for commercial development, in particular, Ob and Parusovoye (Fig. 3.14).

The history of the company began with the development of the Yamburg oil and gas condensate field. In terms of the volume of proven initial reserves – 6.9 trillion cubic meters of gas – it ranks fifth in the world.

In the future, it is planned to supply hydrocarbon raw materials to the released production capacities of the field from the gas-bearing areas adjacent to Yamburg. Among them are deposits in the waters of the Ob and Taz bays.

The Zapolyarnoye oil and gas condensate field is one of the largest in the world. Its initial reserves are more than 3.5 trillion cubic meters of gas, about 60 million tons of gas condensate, and 20 million tons of oil. Commercial exploitation of the deposit began in 2001 with the development of the Cenomanian deposit. Currently, the extraction of hydrocarbon raw materials is carried out at five integrated gas treatment plants (IGTP) – three Cenomanian and two Valanginian.

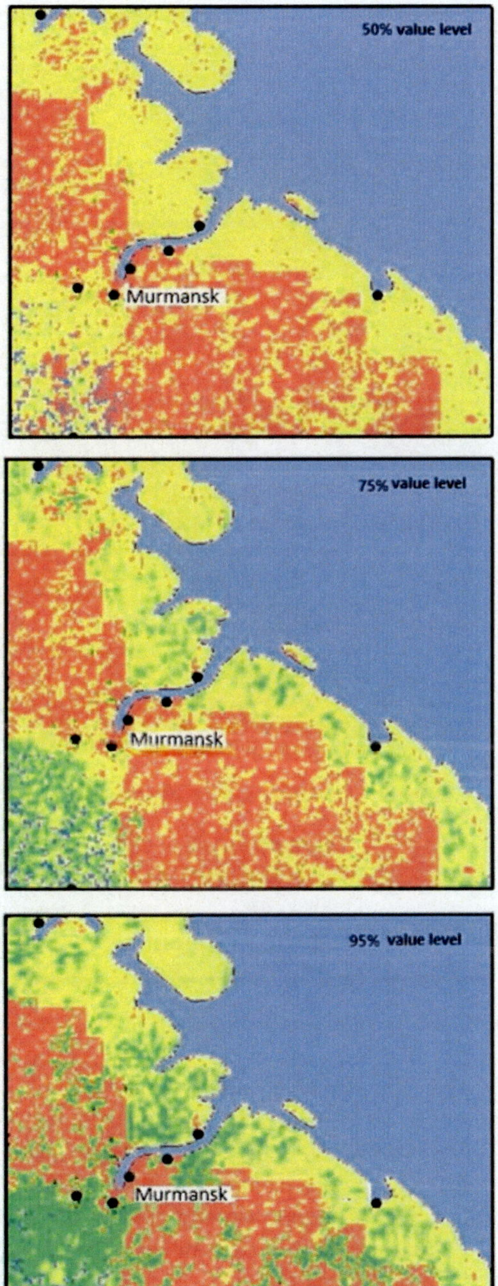

Fig. 3.13 Assessment of the stability of the ecosystems of the Kola Peninsula coast to technogenic NOx emissions: eutrophication effects. Black dots represent different cities

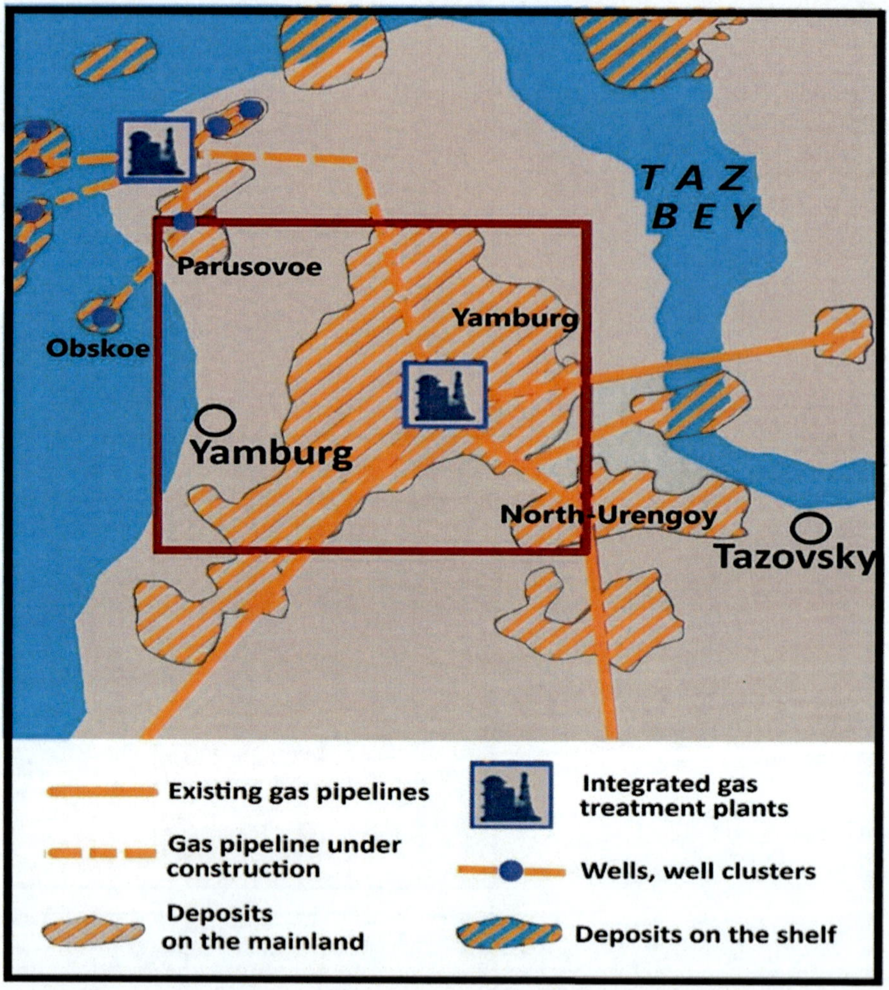

Fig. 3.14 The main gas condensate fields of Gazprom Dobycha Yamburg LLC

There are more than 2400 gas and gas condensate wells in the well fund of the enterprise (Fig. 3.15). The total length of the operated gas pipelines-loops exceeds 2200 km.

Gas is delivered to central Russia and European countries via main gas pipelines after the necessary technological preparation. One of the integrated gas treatment plants, IGTP-5 in the Yamburg field is shown in Fig. 3.16.

As already noted, Gazprom Dobycha Yamburg LLC is the leading mining enterprise of Gazprom PJSC, accounting for about 40% of all natural gas and condensate production. This enterprise uses the most advanced technologies of gas production and treatment. The company's motto is sustainable development, combining production activities with minimizing environmental impact (Arabsky & Bashkin, 2015; Arabsky et al., 2015; Arno et al., 2021).

Fig. 3.15 Well cluster of the Yamburg field

Fig. 3.16 IGTP-5 of the Yamburg field

However, the complex anthropogenic impact on the ecosystems of the Far North, inevitably produced during the extraction of hydrocarbon raw materials, requires regular and objective assessment, which makes it possible to monitor the dynamics of changes in the environment with a high degree of probability. It is most expedient to carry out a retrospective and forecast assessment based on reasonable criteria, which will improve the efficiency of environmental management at the production facilities of Gazprom Dobycha Yamburg LLC. It is also significant that the study of the response of the natural environment to the same type of impact makes it possible to apply the method of analogies for other deposits of the Far North (Bashkin et al., 2013; Bashkin & Galiulin, 2015).

It should be noted that the main anthropogenic emissions at the enterprise are nitrogen oxides (NO_x) and greenhouse gases (carbon dioxide – CO_2, methane – CH_4, and nitrous oxide – N_2O, as a product of the transformation of nitrogen oxides) (Arabsky et al., 2016). Nitrogen oxides have both acidifying and eutrophic effects on ecosystems in the zone of impact of the enterprise and have a corresponding impact on the development of the geoecological situation. The degree of their impact in international practice is assessed using indicators of critical loads as a basic criterion for calculating geoecological risks (Bashkin, 2006a, b). The assessment of greenhouse gas emissions has both an environmental and an economic component, which is directly related to the principles of sustainable development declared in Gazprom Dobycha Yamburg LLC (Markelov et al., 2013).

Thus, the purpose of this section is to develop reasonable criteria for assessing geoecological risks in the impact zones of Gazprom Dobycha Yamburg facilities using international approaches. To achieve these goals, the following tasks were set:

- Development of a system for assessing and monitoring the environmental situation in connection with atmospheric emissions of pollutants during gas production in the Far North
- Quantitative assessment of geoecological risks and their excess in the area of activity of Gazprom Dobycha Yamburg LLC

3.2.1 *Development of a System for Assessing and Monitoring the Geoecological Situation in Connection with Atmospheric Emissions of Pollutants During Gas Production in the Far North*

Currently, the priority of environmental protection measures is determined based on the ranking and classification of pollution sources according to the hazard class of pollutants and the volume of their emissions. The maximum possible reduction of pollutant emissions in the sequence of their hazard class is envisaged (Table 3.5).

Based on the data in Table 3.5 and the existing structure of emissions in the Russian gas industry, nitrogen oxides can be attributed to the number of priority pollutants. The definition of the NO_x hazard class is based on sanitary and hygienic standards that take into account the effects of their inhalation effects on human health. However, the negative impact of NO_x on biota is also possible at concentrations below the MPC for humans, which is associated with the effects arising from medium- and long-term anthropogenic loads as a result of the deposition of

Table 3.5 Hazard classes of pollutants present in process plant emissions

Name of the pollutant	Pollutant hazard class
Nitrogen oxides – NO_x (NO, NO_2)	3
Soot (benz (a)pyrene)	3(1)
Sulfur dioxide (SO_2)	3
Carbon monoxide (CO)	4
Methane (CH_4)	4

pollutants from the atmosphere and accumulation in soils. The ecological danger of "excess nitrogen" for polar and tundra ecosystems is associated with the effects of their acidification and eutrophication (Bashkin, 2016, 2017a, b, 2018).

Quantitative parameterization of exposure is associated with the assessment of exposure levels from stationary point or linear sources of atmospheric emissions of pollutants. It can be performed using special industry techniques developed in the gas industry, or mathematical modeling of the dispersion and migration of pollutants in conjugate media (the surface layer of the atmosphere, soils, soil-ground, and surface waters). These models take into account such indicators as the height of the source and the average intensity of pollutant emissions, their duration (exposure time), and meteorological characteristics of the area where the source of emissions is located (wind direction and speed, precipitation intensity, etc.).

The results of modeling the processes of dispersion of gaseous emissions of pollutants from a stationary source in the surface layer of the atmosphere are usually presented in the form of diagrams (cartographies), which depict the fields of deposition of pollutants of various intensities on an appropriate scale, characterizing the total area of exposure (Fig. 3.17). It can be seen that these precipitation slightly exceed 1 kg/ha/year, which is significantly less than the background level (approximately 3.5 kg/ha/year).

The dispersion and deposition of pollutants (NO_x) from the integrated gas treatment plant (IGTP-2) commissioned at the Yamburg gas condensate field located within the main watershed have also been simulated, which affects the features of atmospheric dispersion of pollutants.

Taking into account the obtained meteorological data and data on NO_x emissions, model calculations of the total precipitation of nitrogen oxides for terrestrial

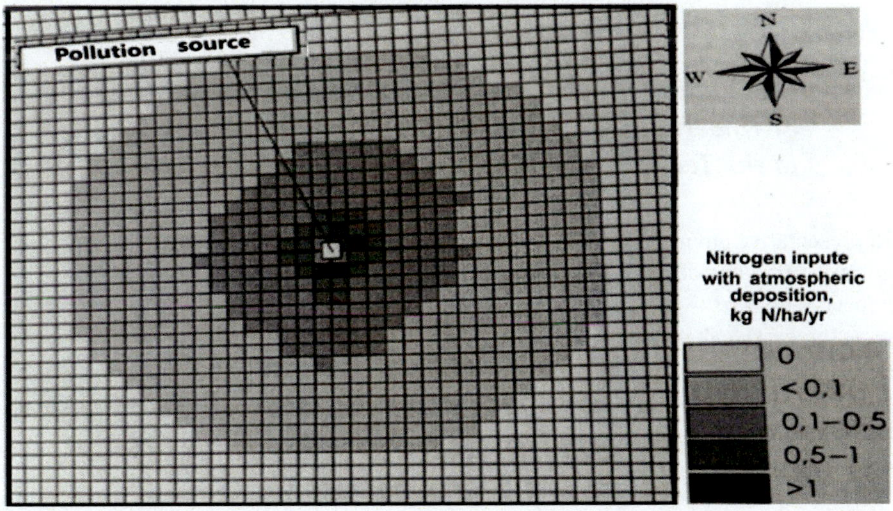

Fig. 3.17 An example of a scheme obtained from the results of modeling nitrogen deposition levels from a chiseled source – the Ural gas pumping unit (the size of the modeling cell is 1 × 1 km)

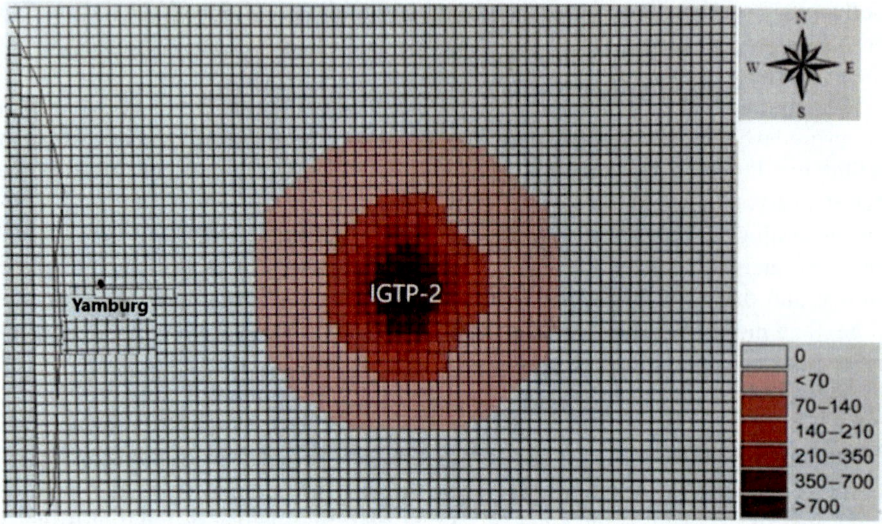

Fig. 3.18 Results of modeling of nitrogen oxide precipitation (according to the data of the Novy Port weather station (g-eq./ha/year))

ecosystems in the IGTP-2 emission zone were carried out. The results obtained are shown in Fig. 3.18. It can be seen that in the area adjacent to the industrial site where IGTP-2 technological facilities are located, nitrogen deposition reaches more than 700 g-eq./ha/year (>20 N/ha/year). Taking into account the wind rose, these precipitation events extend approximately 15–20 km from the source of emissions. These data were used to assess geoecological risks.

3.2.2 Calculation of Geoecological Risk Values and Estimation of Uncertainty in the Emission of Pollutants in the Territories of Gas Condensate Fields

Representative environmentally sound criteria for assessing the geoecological situation in the considered coastal Arctic regions include parameters (indicators) of geoecological risks for different scenarios of emission impacts (retrospective, current, forecast), calculated using the methodology of critical loads.

Modern quantitative methods of assessing geoecological risks require the use of a significant amount of structurally heterogeneous information, which necessitates the development of special software for converting such data into a single format within the GIS project. In addition, the geoecological risk assessment algorithm includes special modeling of pollutant deposition levels for the "technogenic emission source – air environment – terrestrial ecosystems" system, probabilistic calculations of parameters of permissible pollutant loads and their exceedances, as well as uncertainty analysis related to the specifics of such information.

The values of permissible loads are correlated with critical loads. The concept of critical load (or CL value), according to the definition (Nilsson & Grennfelt, 1988), corresponds to the maximum level of pollutants entering the terrestrial (or aquatic) ecosystem, which is not accompanied by irreversible changes in the functioning of the biota, the structure of the ecosystem, and its productivity over a long period of time – up to 100 years or more. Thus, the assessment of CL values presupposes the determination of a threshold for the entry of pollutants into ecosystems, after exceeding which negative consequences for living organisms and the ecosystem as a whole are possible, while no violations and adverse effects are observed below this level. Unlike traditional "environmental" indicators for Russia and most other countries that normalize concentrations of pollutants in individual environments, the value of CL is an ecosystem indicator.

3.2.2.1 Algorithm for Calculating the Values of Critical Loads for Terrestrial Ecosystems of Technogenic Compounds of Acidifying and Eutrophying Action

Calculations of CL values include parametrization of the main migration flows of elements specific to different bioclimatic and landscape conditions. A detailed methodology for calculating CL values is available on the website of the Working Group on the Calculation and Mapping of CL of the Convention on Transboundary Air Pollution (www.icpmapping.org) and in the Bashkin & Priputina's monograph (2010). Let us focus on some of the fundamental provisions used in these calculations. Most often, simple biogeochemical models are used to quantify CL values, which include two types of equations:

- Equilibrium equations of a simple balance of the masses of elements and (or) their compounds in the soil (considered soil layer)
- Equations describing the intensity of the main biogeochemical flows of elements in ecosystems, taking into account the peculiarities of their formation

In these models, a number of assumptions are made, namely:

- The depth of the soil layer under consideration is conditionally equal to the depth of the root zone, which makes it possible to neglect the nutrient cycle.
- Evapotranspiration occurs on the surface of the soil profile.
- Precipitation moisture seepage is constant throughout the soil profile and occurs only vertically.
- Physico-chemical constants are assumed to be homogeneous throughout the soil profile.
- Internal flows of elements (nitrogen fixation, etc.) do not depend on the chemical conditions of the soil (such as pH).

In addition, internal ecosystem interactions and processes such as intraspecific competition or the presence of pests, removal of elements from the soil with an increase in the part of terrestrial biomass that annually returns to the surface with the fall of vegetation, and others are not taken into account. Since these models describe conditions relative to the equilibrium biogeochemical state, they require the use of long-term averaged values of incoming flows as input information. Seasonal, interannual, and other short-term dynamic changes of indicators in these models, as a rule, are not taken into account In Europe, where in most countries there is a detailed network of in situ monitoring observations, measured averages are usually used to calculate the values of CL. For example, in the Netherlands, the detail of the network of soil-geochemical testing of agricultural and forest lands is 500x500 m, and for each point in the nodes of this grid, pH values, granulometric composition of soils, organic matter content, etc., were measured with varying frequency. However, when calculating CL for relatively large-scale regions, regions with a weak degree of knowledge, or a complex structure of natural organization, problems arise associated with a high degree of uncertainty and differentiation of input parameters. In this case, in our opinion, the most correct estimates can be obtained based on the use of probabilistic calculation methods (for example, the Monte Carlo method), when a possible range of values of CL is calculated for different sites, followed by an analysis of indicators corresponding to different levels of values (as a rule, 25%, 50%, 75%, 95%). This approach also has advantages in terms of including the data obtained in the probability assessment of geoecological risks and subsequent management decisions. We have implemented this approach on the example of a number of GAZPROM facilities (Bashkin et al. 2002a; Bashkin & Priputina, 2010).

The purpose of calculating the values of critical loads of acid-forming compounds (or CL acidity) is to quantify the potential of ecosystems in terms of neutralizing the acid component of atmospheric precipitation, which depends on the balance of chemically active protons and cations in natural systems. It is known that the main agent of the acidity of precipitation is sulfur oxides, with the excess emission of which in the 1960s to 1980s of the twentieth century was associated with the magnitude of the problem of acid rain in Central and Northern Europe. Therefore, in the framework of studies under the Convention on Long-Range Transboundary Air Pollution in 1995–1997, it was decided to use the term "critical load of maximum sulfur" ($CL(S)_{max}$) to assess the permissible parameters of acid-forming compounds entering ecosystems. It is believed that this value corresponds to a safe level of precipitation of sulfur oxides in the absence of other acidic compounds in atmospheric precipitation. In the case of the presence of other acid-forming compounds in the air, primarily nitrogen oxides, it is necessary to correct (reduce) the permissible sulfur intake (Fig. 3.19). In recent years, after a noticeable reduction in SO_x emissions in Europe, in many industrial regions, including in the Russian Federation, nitrogen oxides have begun to play a significant role in the formation of acid precipitation, reducing atmospheric emissions of which is a more difficult practical task.

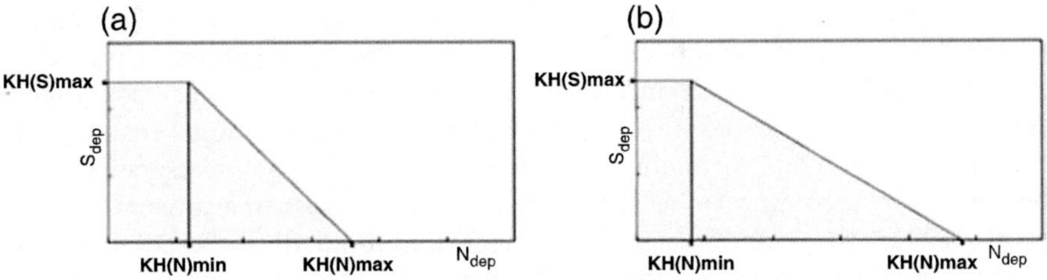

Fig. 3.19 Schemes for correcting the maximum sulfur and acidifying nitrogen KN, determined by the calculated values of CL $(S)_{max}$, CL $(N)_{min}$, and CL $(N)_{max,}$ where (**a**) is at a constant denitrification value (the angle of inclination of the function is 45°) and (**b**) is at denitrification parameters depending on nitrogen intake with atmospheric precipitation (a more gentle slope of the function determined by the value f_{de})

The gray zone corresponds to the "pairs" of precipitation levels S (S_{dep}) and N (N_{dep}), which provide conditions under which soil acidity will not be below acceptable critical pH values.

The basic algorithm for calculating the values of CL acidity is based on the use of the law of equivalents:

$$\mathrm{CL}(S)_{\max} = \mathrm{BC}_{\mathrm{dep}} + \mathrm{BC}_{\mathrm{w}} - \mathrm{Cl}_{\mathrm{dep}} - \mathrm{BC}_{\mathrm{up}} - \mathrm{ANC}_{\mathrm{le(crit)}}$$

where, BC_{dep} is the input of Ca, Mg, K, Na cations into the ecosystem with atmospheric precipitation; BC_w is the intra-soil weathering of Ca, Mg, K, Na cations; Cl_{dep} is the entry of Cl anions with atmospheric precipitation; Bc_{upt} is the removal of Ca, Mg, K cations from soils by vegetation due to root nutrition; and ANC $_{le(crit)}$ is the critical leaching of alkalinity. The units of measurement of the values of all parameters of this equation are gram equivalent per hectare per year (g-eq./ha per year).

For nitrogen compounds, several variants of critical loads are calculated, which is necessary to quantify the ratio between the permissible levels of nitrogen and sulfur oxides when they are jointly present in atmospheric precipitation.

The minimum nitrogen load characterizes the lowest level of nitrogen intake, which ensures the preservation of productivity of the ecosystems under consideration. The critical load of minimum nitrogen (CL$(N)_{min}$) is calculated according to the following equation:

$$\mathrm{CL}(N)_{\min} = N_{\mathrm{im}} + N_{\mathrm{upt}}$$

where, N_{im} is the amount of nitrogen annually fixed (immobilized) in the soil due to the processes of creating soil organic matter (g-eq./ha per year) and N_{upt} is nitrogen accumulated in the growth of vegetation biomass (g-eq./ha per year).

Long-term immobilization of nitrogen (N_{im}) is determined by its accumulation as part of a stable pool of soil organic matter. This nitrogen immobilization should not lead to a change in the *C*/*N* ratio compared to similar background soils. The existing expert estimates of annual nitrogen immobilization in European forest soils differ in determining the values of N_{im}, especially for boreal soils. According to the data obtained by Swedish researchers (Rosén et al., 1992), for Scandinavian forests, the long-term dynamics of nitrogen deposition in the soil is 0.2–0.5 kg N/ha per year, which, apparently, is also true for boreal forests of the European territory of Russia, but not for its Arctic regions. The recommendations of the Coordination Center for the Effects of the Convention on Long-Range Transboundary Air Pollution recommend using a value of 1 kg N/ha/year for all types of ecosystems (UBA, 2004), which, apparently, is not entirely correct. Nitrogen removal with biomass growth is estimated as the product of the annual production of the part of biomass that can be removed from ecosystems and the concentration of nitrogen in the considered part of the phytomass.

The critical load on "nutrient" nitrogen allows, on the contrary, to quantify the supply of nitrogen that does not cause its excess in ecosystems (eutrophication), which may be the reason for the change in the species diversity of biocenoses. The critical load of nutrient nitrogen ($CL(N)_{nutr}$) is calculated according to the following equation:

$$CL(N)_{mitr} = CL(N)_{min} + N_{le(acc)} / (1 - f_{de})$$

where, CL $(N)_{min}$ is the value of the critical load of minimum nitrogen; $N_{le(acc)}$ is the permissible leaching of nitrogen from soils into soil-groundwater; and f_{de} is the denitrification coefficient.

The f_{de} coefficient depends on the intensity of denitrification processes in soils. For soils characterized by good drainage conditions, the value $f_{de} = 0.1$ is used, and for soils of heavy granulometric composition with difficult drainage conditions, the value $f_{de} = 0.4$–0.8 is used.

The maximum nitrogen load determines the permissible parameters of nitrogen oxides entering ecosystems, at which there is no drop in the level of acidity (pH values) below the critical level and, at the same time, the nutrient regime of soils favorable for biota is maintained. The value of the critical load of maximum nitrogen ($CL(N)_{max}$) is calculated according to the following equation:

$$CL(N)_{max} = CL(N)_{min} + CL(S)_{max} / (1 - f_{de})$$

The parameters included in this equation and their calculation methods are discussed above.

3.2.2.2 Algorithm for Assessing Geoecological Risks Using the Values of Critical Loads

Within the framework of studies under the Convention on Long-range Transboundary Air Pollution, environmental risk assessments for terrestrial and freshwater ecosystems are carried out in connection with atmospheric air pollution, using the calculation of so-called "critical load excess" (Ex(CL)) to determine the probability of environmental violations. The excess of CL is estimated as the difference between the values of the actual (or predicted) pollutants entering ecosystems with atmospheric precipitation and the corresponding values of CL (Hall et al., 2003). According to this approach, positive values of excess CL (Ex(CL) > 0) mean that the permissible levels of exposure to pollutants for specific ecosystems (territories) have been exceeded and environmental disturbances in the structure and functions of natural ecosystems caused by exposure to the pollutant in question are possible in them. Since calculations of CL and precipitation of pollutants under the Convention are performed using deterministic data (based on average values), the results obtained are not "classical" risk indicators.

In our opinion, from a methodological point of view, more correct estimates of geoecological risks can be obtained by combining this approach with probabilistic methods, when the risk characteristic includes the calculation of the probability of exceeding CL (or the probability of exceeding permissible exposure levels), which can be expressed by the following formula:

$$\mathrm{Risk}(X) = P(\mathrm{Ex}(\mathrm{CL}) > 0) = P([X]_{\mathrm{dep}} - \mathrm{CL}(X) > 0), \tag{3.1}$$

where, *Risk(X)* is the geoecological risk to ecosystems associated with exposure to a specific pollutant (X), *CL(X)* is the permissible level of this pollutant entering the ecosystem (critical load), $[X]_{dep}$ is the existing or predicted level of pollutant intake into the ecosystem with atmospheric precipitation, *Ex(CL)* is exceeding the acceptable level of admission (*CL*), and *P(Ex(CL)* is the probability of exceeding the permissible level (*CL*).

Technically, the calculation of the probability of exceeding CL can be performed, for example, using the Monte Carlo method (Sobol, 1985). In accordance with the proposed formula, the excess of CL reflects the ratio between the magnitude of the recipient's exposure (within the boundaries of the zone of influence of an existing or projected production facility) and the level of exposure that is safe for this recipient. Since CL is not an integral indicator of ecosystem sustainability, but is focused on justifying permissible impacts with respect to certain pollutants and specific effects of exposure, this allows using this indicator to quantify partial geoecological risks in the same way as it is done in the calculations of multifactorial technogenic risks or risks to human health (Onishchenko et al., 2002; Kolesnikov, 2008). It should be recognized that in this form, the proposed formula does not include quantitative parameters of environmental damage associated with economic activity. The assessment of environmental damage, including industrial enterprises and infrastructure, in this case is based on the qualitative characteristics of those changes that are possible in specific ecosystems when the permissible levels of pollutants entering them are exceeded, which is taken into account in the calculations of CL.

If, on the basis of experimental studies, appropriate quantitative dependences are determined between the level of the considered impact and the environmental changes associated with it, then the necessary coefficients (or values) can be included in the formula proposed by us, allowing to take into account the magnitude of environmental damage with a given probability of an adverse event. In addition, as will be shown below on the examples of the implementation of this approach, if such assessments are carried out on a regional scale – for different types of ecosystems within the entire impact zone (from the maximum level of pollutant receipts to the minimum) – then the resulting spatial picture of the distribution of zones with different levels of probability of excess of CL is a quantitative characteristic of territorial risk (Bashkin & Demidova, 2008; Bashkin, 2017b, 2018).

3.2.2.3 Calculation of Critical Loads – Permissible Levels of Nitrogen Supply to Terrestrial Ecosystems

The estimation of CL values of eutrophying and acidifying nitrogen compounds for the Arctic ecosystems of the Yamburg model site was carried out in accordance with a generally accepted algorithm (UBA, 2004; Bashkin, 2006a, b, 2009; Bashkin & Priputina, 2010). In the calculations performed using probabilistic methods, data from attribute tables of a special GIS project created were used. For the calculation algorithm, a set of parameters assigned variable was determined. Modeling of CL values and their exceedances based on an array of input data was carried out using the Monte Carlo method ($N = 1000$). Based on the results of model estimates from a set of 1000 probable values, CL values were obtained for each 1 × 1 km modeling cell in the GIS project attribute database (separately for the effects of eutrophication and acidification of ecosystems). The spatial distribution of the obtained values was reflected in the form of separate thematic layers of the GIS project. The results of the calculations are presented in Figs. 3.20, 3.21, 3.22, 3.23, 3.24, 3.25, 3.26, and 3.27.

3.2.2.4 Assessment of Exceedances of Critical Loads – Values of Geoecological Risk

To characterize the geoecological risk, the following calculations were carried out:

- Calculation of probabilistic excess CL of eutrophying nitrogen compounds (EX(N))
- Calculation of probabilistic excess CL of acidifying nitrogen compounds (EX(S))
- Calculation of the probabilistic excess of sulfur and nitrogen CL (EX($N + N$))

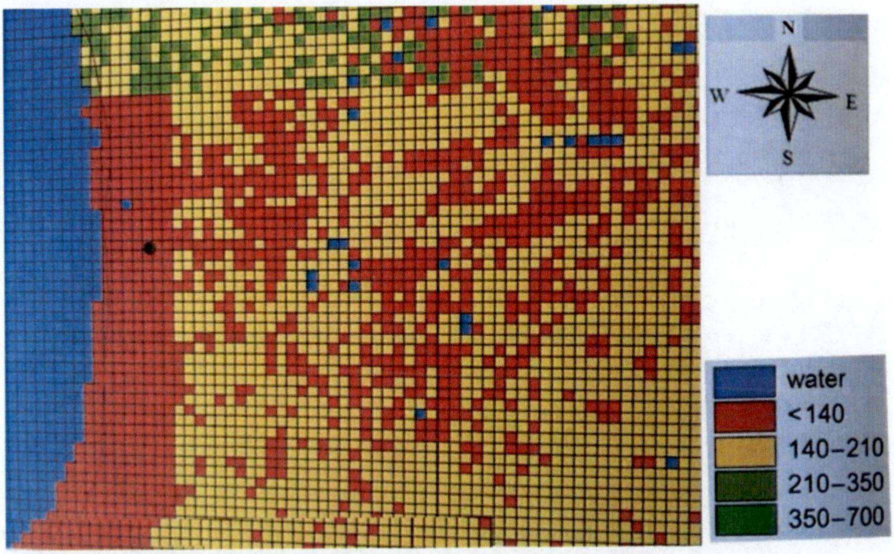

Fig. 3.20 CL of eutrophying nitrogen compounds, 25% level of values (g-eq./ha/year)

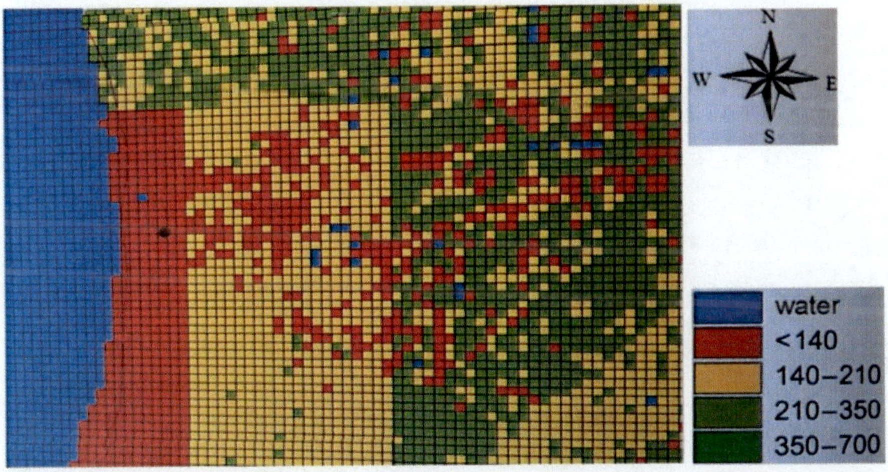

Fig. 3.21 CL of eutrophying nitrogen compounds, 50% level of values (g-eq./ha/year)

The probability of exceedances was modeled using the Monte Carlo method ($N = 1000$). As probable values of nitrogen deposition levels, the results obtained by modeling NO_x dispersion from a point source of emissions – IGTP-2 (Fig. 3.18) – were used.

The obtained data on the probability of exceeding the permissible parameters of exposure to eutrophying and acidifying nitrogen compounds for terrestrial ecosystems of the Yamburg model area (expressed in the probability units) were ranked in

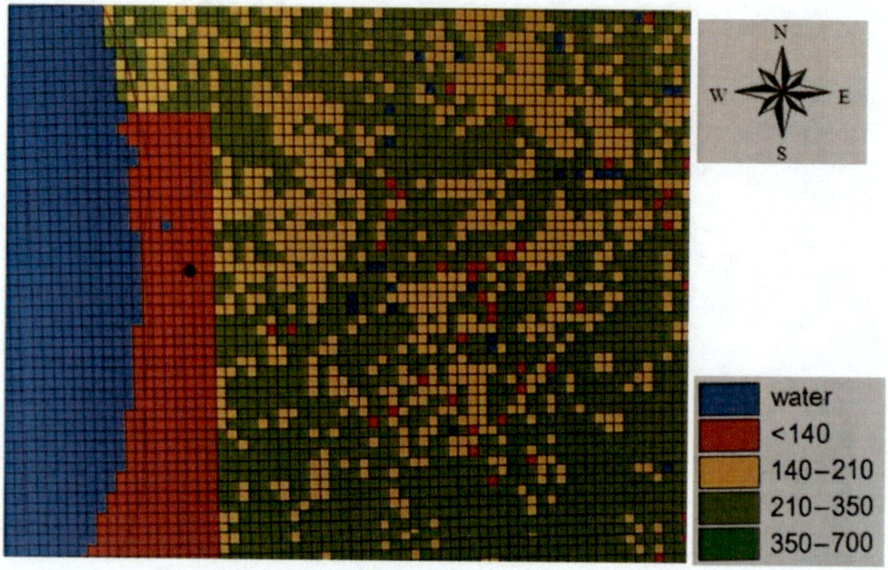

Fig. 3.22 CL of eutrophying nitrogen compounds, 75% level of values (g-eq./ha/year)

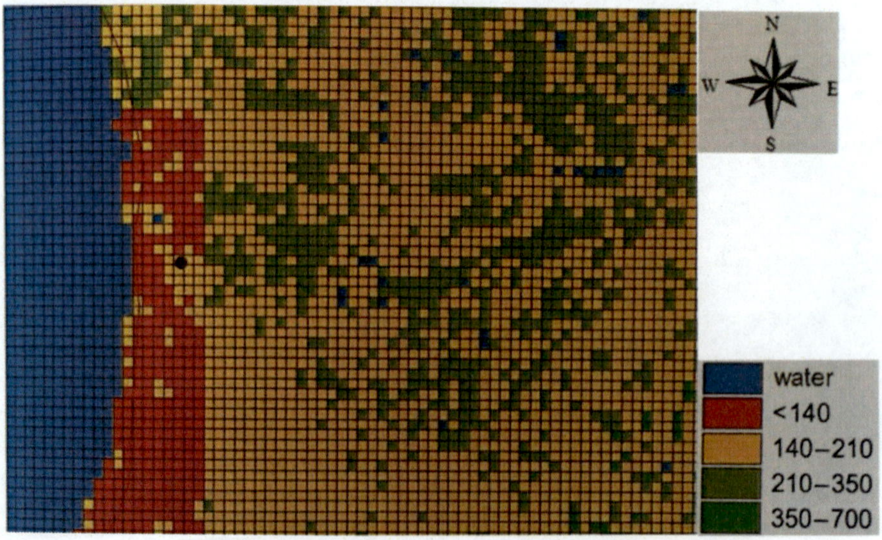

Fig. 3.23 CL of eutrophying nitrogen compounds, 95% level of values (g-eq./ha/year)

accordance with existing approaches into 5 classes: less 0,05; 0,05-0,25; 0,25-0,50; 0,50-0,75; 0,75-0,95; more than 0.95. The results of the calculations are presented in the form of a series of schematic maps and thematic layers of the GIS project, which provides additional opportunities for analysis and interpretation of the data obtained (Figs. 3.28, 3.29, and 3.30).

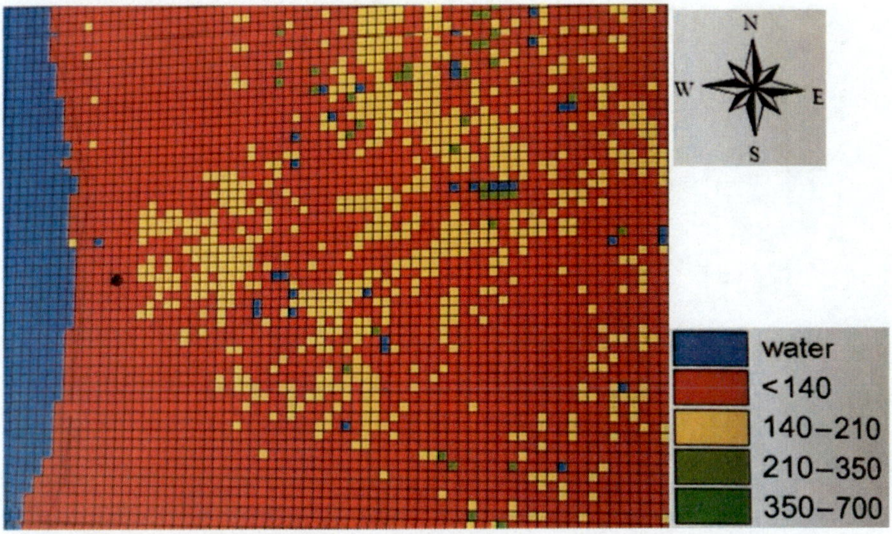

Fig. 3.24 CL of acidifying nitrogen compounds, 25% level of values (g-eq./ha/year)

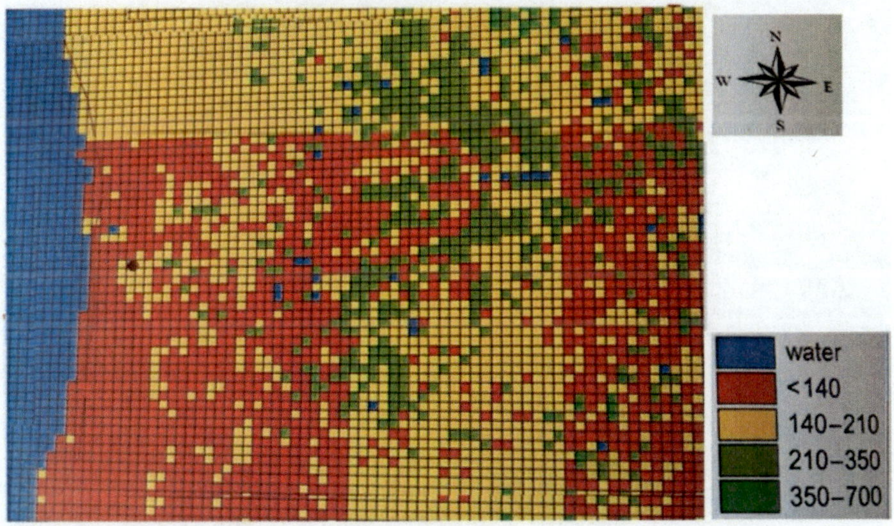

Fig. 3.25 CL of acidifying nitrogen compounds, 50% level of values (g-eq./ha/year)

The consequence of the excess of CL eutrophying effects of nitrogen, shown in Fig. 3.28, is the change of natural moss-lichen vegetation in ecosystems around IGTP-2 to reed grass-meadow, as shown in Fig. 3.29.

At the maximum level of emissions, the excess of $CL(N)_{nut}$ is significant for almost the entire 30-kilometer impact zone, which is associated with a change in the

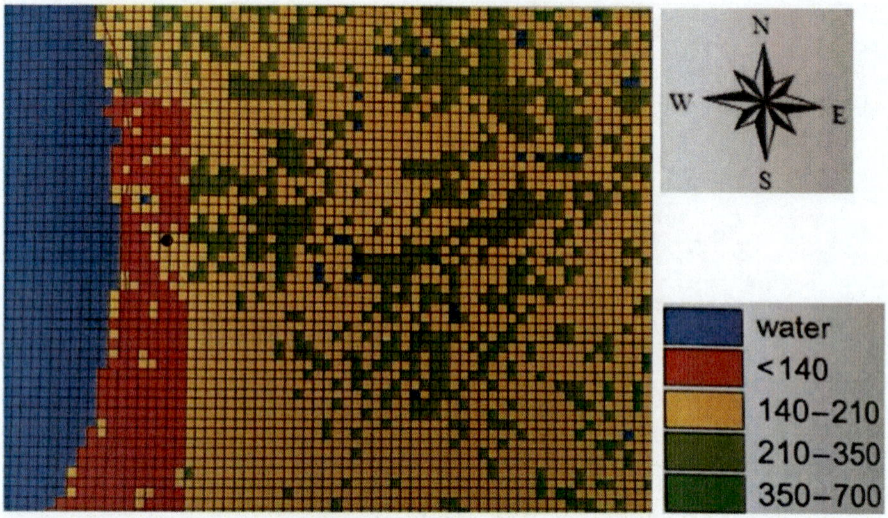

Fig. 3.26 CL of acidifying nitrogen compounds, 75% level of values (g-eq./ha/year)

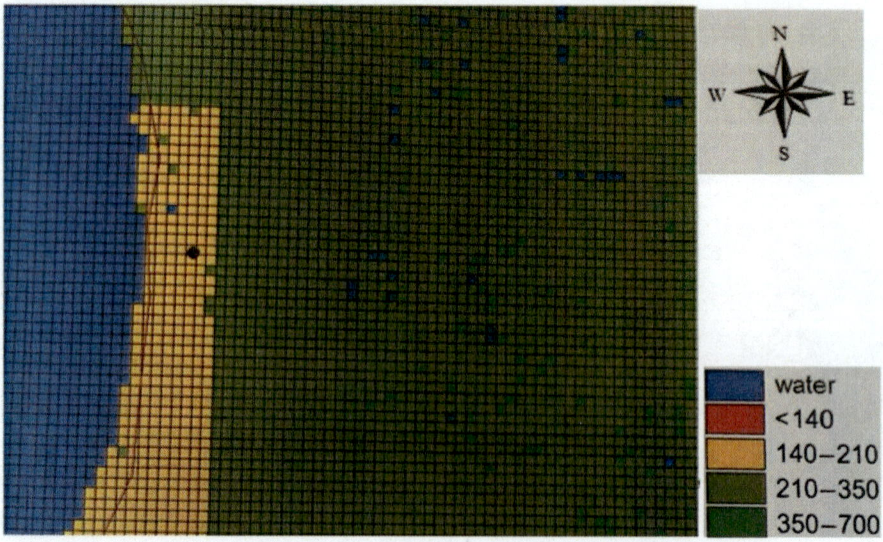

Fig. 3.27 CL of acidifying nitrogen compounds, 95% level of values (g-eq./ha/year)

structure of the vegetation cover of this territory due to the death of lichen species and mosses, as well as the disappearance of some oligotrophic vascular species. The most likely effect of eutrophication of ecosystems in this area is an increase in the number of sedges and cereals and an increase in the overall productivity of

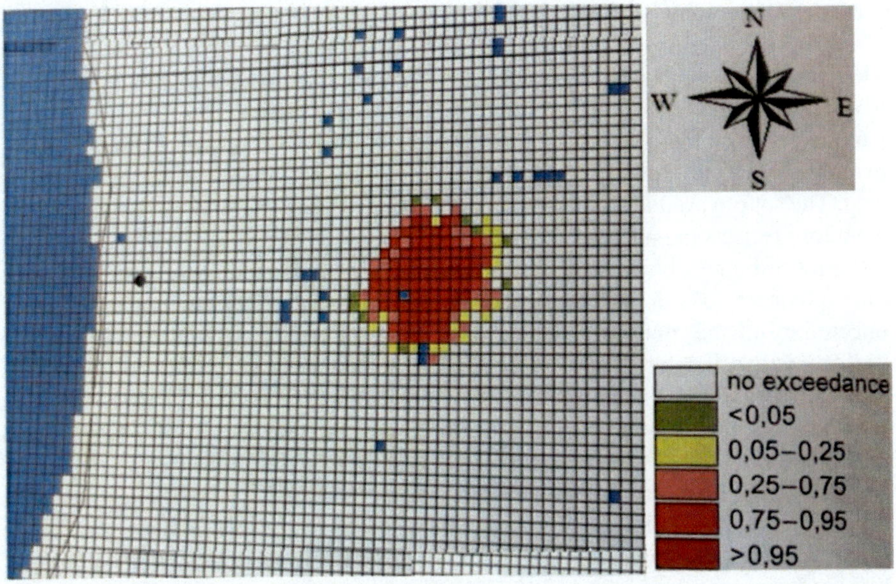

Fig. 3.28 The probability of exceeding the permissible parameters of exposure to eutrophying nitrogen compounds

Fig. 3.29 Reed grass meadow near IGTP-2

phytocenoses, which can lead to a change in the thermal characteristics of the soil and vegetation layer.

At the same time, the excess of critical loads in acidity (Fig. 3.30) is of less importance, since tundra soils have a significant buffer potential for acidity due to their saturation with aluminum ions.

3.2.2.5 Uncertainty Analysis

Modeling of CL values and their exceedances using the Monte Carlo method allows us to take into account many problems associated with the unreliability of input parameters (Risk and Uncertainty, 1997) and reduce the existing complexity, spatial heterogeneity, and temporal dynamics of most natural characteristics of the study area. The set of variable values for each of the input parameters of the mass balance equations reflects the most probable range (interval) of indicators characterizing the intensity of biogeochemical, soil, and landscape-geochemical processes in the tundra ecosystems of the Far North. Thus, it was possible to significantly reduce the uncertainty of the input parameters in comparison with the results obtained with deterministic estimates.

Denitrification (f_{de}) and removal of elements with phytomass because of reindeer grazing (Y_{hpp}) should be attributed to the parameters whose contribution to the overall uncertainty of the estimates made regarding the effects of eutrophication may be significant. In the case of denitrification, there is practically no experimental data reflecting the possible intensity and spatial heterogeneity of this process in the types of ecosystems under consideration. Estimates of the removal of elements with phytomass during reindeer grazing are difficult due to the lack of reliable statistics on this type of economic activity.

The uncertainty that arises when modeling the levels of nitrogen deposition entering ecosystems because of emission, dispersion, physico-chemical transformation, and deposition of nitrogen compounds on the Earth's surface is primarily

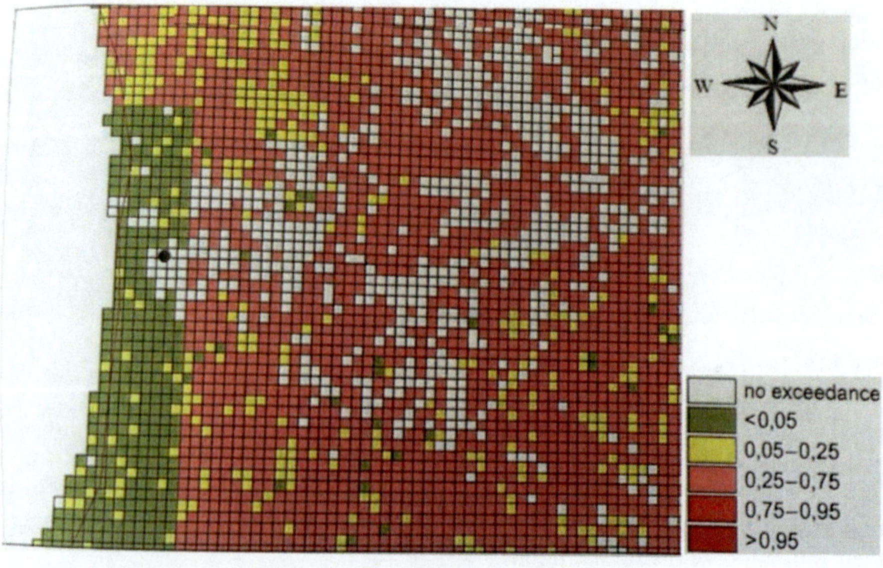

Fig. 3.30 The probability of exceeding the permissible parameters of the impact of acidifying nitrogen compounds

related to the stochasticity of many processes occurring in the atmosphere. The uncertainty was reduced by using in the calculations of exceedances of critical loads variants of nitrogen deposition values calculated from meteorological indicators of several stations located in the considered gas production region in Yamal-Nenets region (YaNAO), Russia.

3.3 Assessment of Geoecological Risks in Impact Zones of Arctic Coastal Ecosystems of the Yamal Peninsula

The relevance and importance of environmental conservation issues in the development of natural resources of the Russian Far North, whose ecosystems are characterized by a reduced potential for resistance to external anthropogenic impacts, are now generally recognized. Currently, the Bovanenkov group of hydrocarbon deposits (BGCF), located in the western part of the Yamal Peninsula (Fig. 3.31) and characterized by large reserves of natural gas and condensate, is among the priority objects for the development of coastal Arctic gas fields.

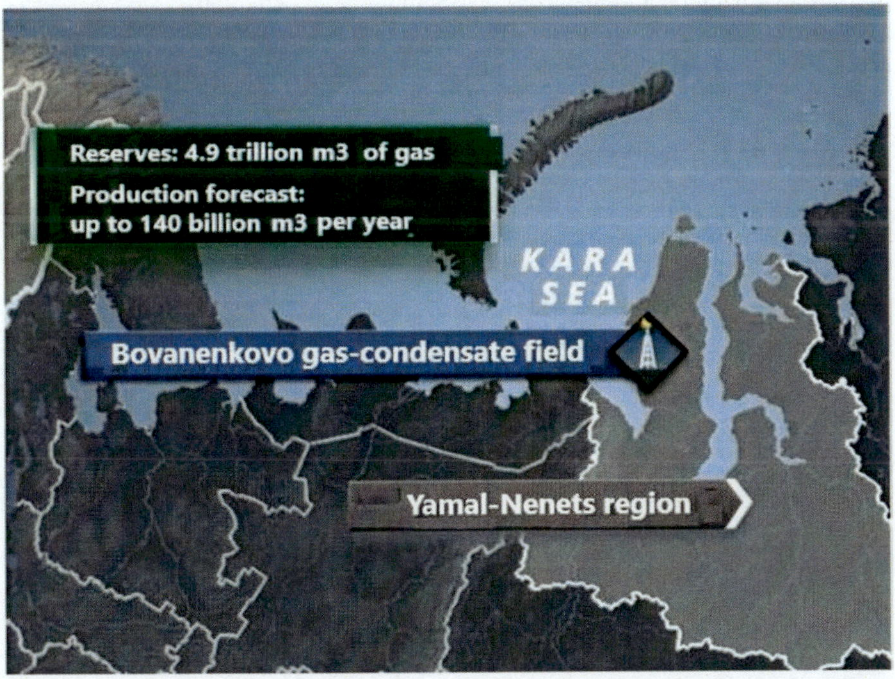

Fig. 3.31 Bovanenkovo gas condensate field, BGCF

3.3.1 Assessment of Geoecological Risks in the Zone of Exposure to NO_x Emissions from the Bovanenko Gas Condensate Field

The assessment of geoecological risks for ecosystems in the zone of impact of BGCF emissions was carried out according to the standard algorithm (Fig. 3.32).

Stage 1: Hazard identification According to the plans for the development of BGCF (Samsonov et al., 2007a), the development of the field is carried out in stages, with the sequential commissioning of technological facilities, which will affect the dynamics of the intensity of pollutant emissions. The overall structure of atmospheric emissions is dominated by methane, carbon monoxide, and nitrogen oxides (Bashkin, 2014; Arabsky et al., 2015). The most significant hazard factor for the regional ecology of Western Yamal will be an increase in the level of anthropogenic nitrogen compounds entering the environment, the excess of which, in addition to direct risks of toxicity to biota due to high concentrations of NO_x in the air, can determine a complex of indirect (secondary) environmental risks associated with eutrophication and acidification (Fig. 3.33).

According to the data of the impact assessment on biota, the species diversity of phytocenoses in the zone of impact of the Bovanenko deposit is determined by the complexity of the combination of vascular species, mosses, and lichens. Cereals, sedge, clove, buttercup, cruciferous, and willow predominate; the composition of mosses is quite diverse. The basis of lichen biota is epigeal species (84%), but there are no conditions for their widespread distribution due to significant waterlogging,

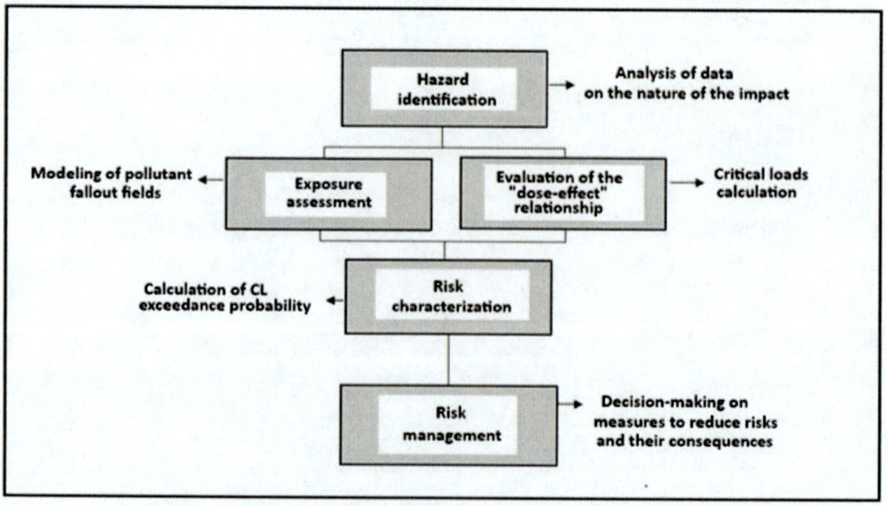

Fig. 3.32 Sequence of ecosystem risk assessment procedures based on the calculation of the probability of CL exceedances

and the constant grazing of deer in previous years led to the fact that forage species of lichens were replaced by little-eaten species, as well as cortical and scale. The bases of the diet when using this territory for deer pastures are cereal species, sedges, shrubby willows, and mosses. As a result of overgrazing, part of the ecosystems of the studied territory has sparse or severely disturbed vegetation cover, which reduces the productive reserves of phytomass. Potential recipients under the conditions of exposure to nitrogen oxide emissions will be terrestrial lichen species, mosses, the main species of vascular plants (including those listed in the Red Book of the Yamalo-Nenets Autonomous Region (1997)), as well as soil algoflora, which accounts for the main pool of biological fixation of atmospheric nitrogen in tundra soils.

Stage 2: evaluation of the dose-response relationship (calculation of CL values) According to the proposed algorithm for assessing geoecological risks, the determination of permissible (reference) doses of nitrogen oxides on phytocenoses in the zone of exposure to BGCF emissions was carried out using probabilistic methods for calculating CL values in relation to the effects of eutrophication and acidification ($CL(N)_{nut}$ and $CL(S)_{max}$). CL calculations were performed with a spatial resolution of 1 × 1 km. The variable parameters of the mass balance equations, as shown above, were the intake of cations and anions with atmospheric precipitation, intra-soil weathering of cations, the fixation of nitrogen and cations in the production of terrestrial phytomass, and nitrogen immobilization in the organic matter of tundra and marsh-tundra soils. For each ecosystem, using the Monte Carlo method, a

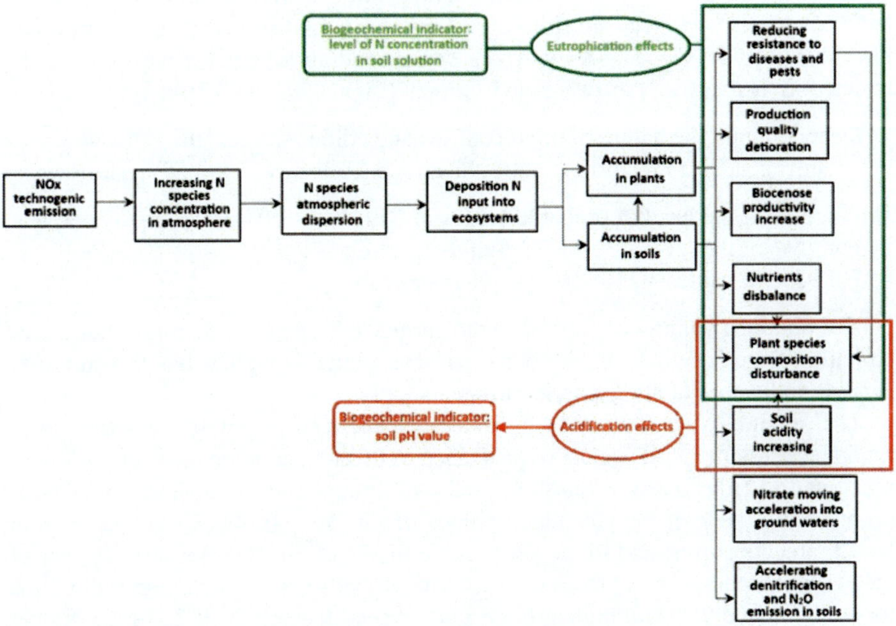

Fig. 3.33 Effects of anthropogenic nitrogen compounds on the environment: events tree

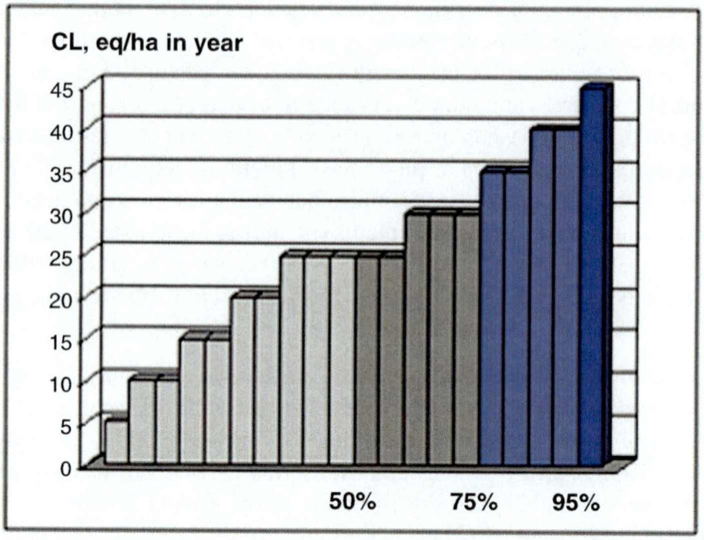

Fig. 3.34 An example of a cumulative distribution curve of possible values of CL for a conditional ecosystem

random sample of 1000 possible CL values was calculated. Subsequent verification showed that such a large number of "runs," with all the randomness of the samples, ensures high repeatability of the results. Then, values corresponding to 50%, 75%, and 95% levels were selected from the obtained cumulative distributions of probable values of CL (Fig. 3.34), for which corresponding schematic maps were constructed in the ArcView software environment (Figs. 3.35 and 3.36).

Depending on the nature of phytocenoses and edifier species, the calculated values of nitrogen CL in relation to eutrophication effects are (Fig. 3.35) as follows:

- For 50% of the level of probable values of $CL(N)_{nut}$ – 210–350 g-eq./ha per year
- For 75% of the level – 210–700 g-eq./ha per year
- For 95% of values – 210–840 g-eq./ha per year

The obtained values correspond to the permissible intake with atmospheric precipitation from 3–5 to 10–12 kg N/ha per year, which is higher than the currently existing indicators of atmospheric nitrogen supply.

The obtained values of $CL(S)_{max}$ reflect the reduced potential characteristic of tundra ecosystems with respect to resistance to the acid component of atmospheric precipitation (Fig. 3.36). According to the minimum estimates (level of 50% of probabilistic values), the calculated values of $CL(S)_{max}$ in the 30 km zone from BGCF objects correspond to the permissible intake of 50–100 g-eq./ha per year of acid-forming compounds. The values of critical loads corresponding to the indicators of 75% and 95% are higher, but do not exceed the values of 280 g-eq./ha per year, which is comparable to the intake of about 3.5–4 kg of sulfur and/or nitrogen

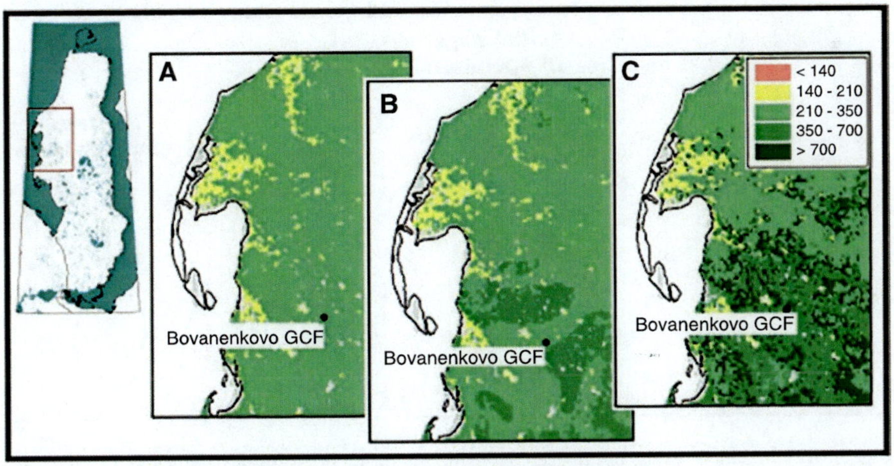

Fig. 3.35 Results of probabilistic calculations of CL of nutrient nitrogen (g-eq./ha per year): (A) 50% level of values; (B) 75% level of values; (C) 95% level of values

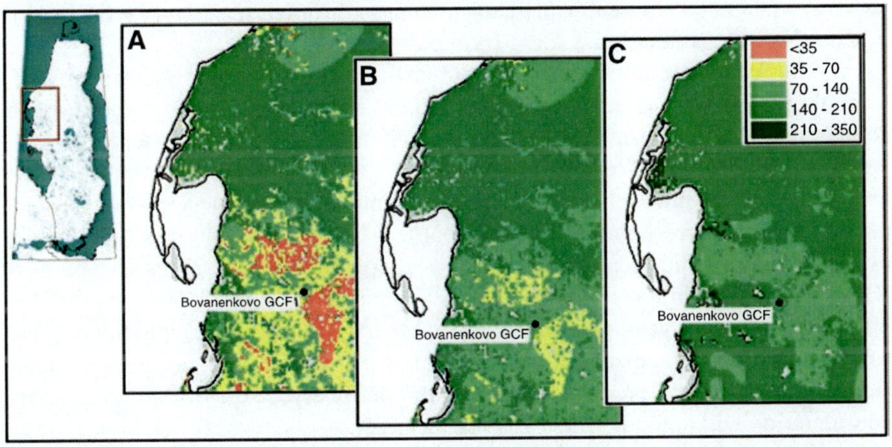

Fig. 3.36 Results of probabilistic calculations of CL of acidifying compounds (g-eq./ha per year): (A) 50% level of values; (B) 75% level of values; (C) 95% level of values

and corresponds to the current values of the total supply of these elements with atmospheric precipitation.

Stage 3: Exposure assessment The scenarios of nitrogen oxide emission by BGCF process units correspond to their phased commissioning according to the field development plans. The assessment of nitrogen input levels in terrestrial ecosystems with atmospheric precipitation was performed using a NO_x scattering model from a point source (IGTP) based on data from total NO and NO_2 emissions. The detail of spatial modeling is 1 × 1 km polygons corresponding to the polygons of CL model calculations. The obtained results are presented in Fig. 3.37.

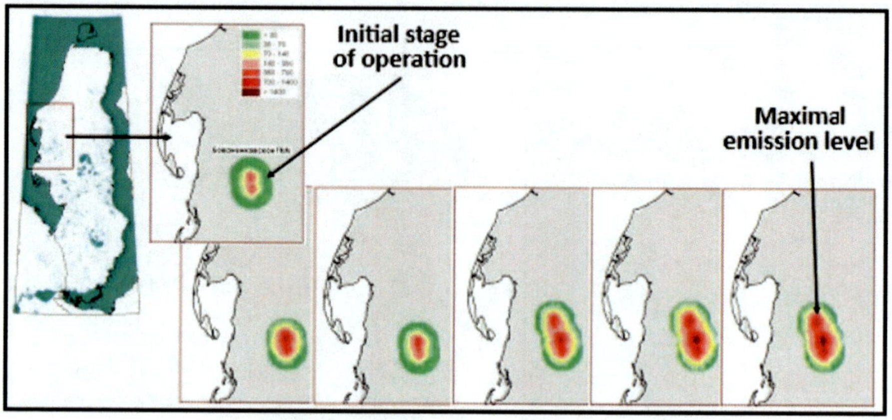

Fig. 3.37 Simulation results of nitrogen deposition at different NOx emission levels

Modeling of the levels of nitrogen input with atmospheric precipitation is performed based on the model of dispersion of atmospheric pollutants from point sources – integrated gas treatment plants (author of the model A. Shvyryaev, Russia). The detail of the simulation is a 1 × 1 km polygon.

As follows from the modeling data, after the phased commissioning of BGCF technological facilities, the regional background of atmospheric supply of nitrogen compounds will increase. The main impact will affect the 30 km zone. In the first years after the commissioning of IGTP-1 and IGTP-2 technological facilities, the level of additional nitrogen deposition for this zone is estimated at 35–140 g-eq./ha per year, which corresponds to the intake of 0.5–2 kg N/ha per year. Subsequently, as the volume of nitrogen oxide emissions increases after the commissioning of IGTP-3, the maximum impact may be more than 2000–2500 g-eq./ha (or 30–35 kg N/ha) per year, which is comparable to the level of precipitation typical for industrial and urbanized areas. However, even with maximum nitrogen emissions, its technogenic precipitation along the periphery of the 30 km zone under consideration will amount to 2–5 kg N/ha per year.

Stage 4: Risk characterization When using probabilistic methods for estimating the values of CL for each ecosystem, we obtain a numerical array of possible values of permissible inputs of pollutants with a different combination of natural factors in accordance with their natural dynamics and variability.

The analysis of the distribution of the obtained values allows us to justify the intensity of anthropogenic loads depending on environmental priorities and economic feasibility. The total distribution of the probable values of CL of nutrient nitrogen (left graph) and CL of acidity (right graph) for the ecosystems of the potential impact zone of BGCF (within 50 km from the emission source) is shown in Fig. 3.38.

The results of calculations of geoecological risks for terrestrial ecosystems in relation to the effects of eutrophication are shown in Fig. 3.39. At all stages, most of the territory under consideration is characterized by the absence of excess $CL(N)_{nut}$

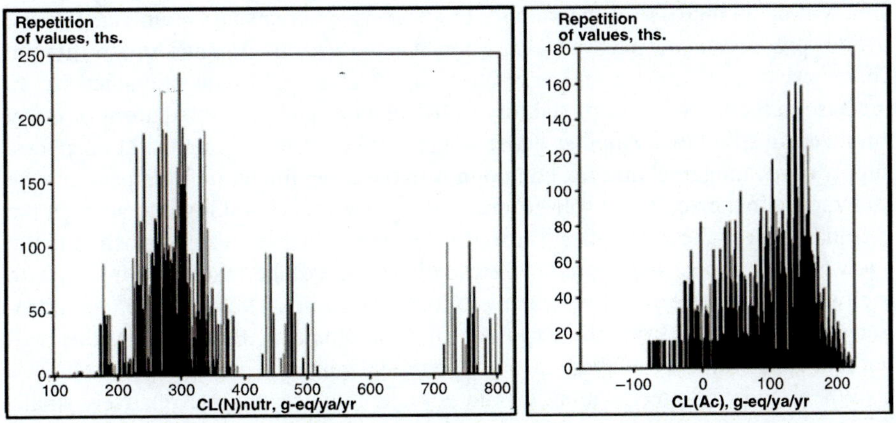

Fig. 3.38 Probabilistic assessment of ecosystem resilience to technogenic NO_x emissions in the BGCF impact zone

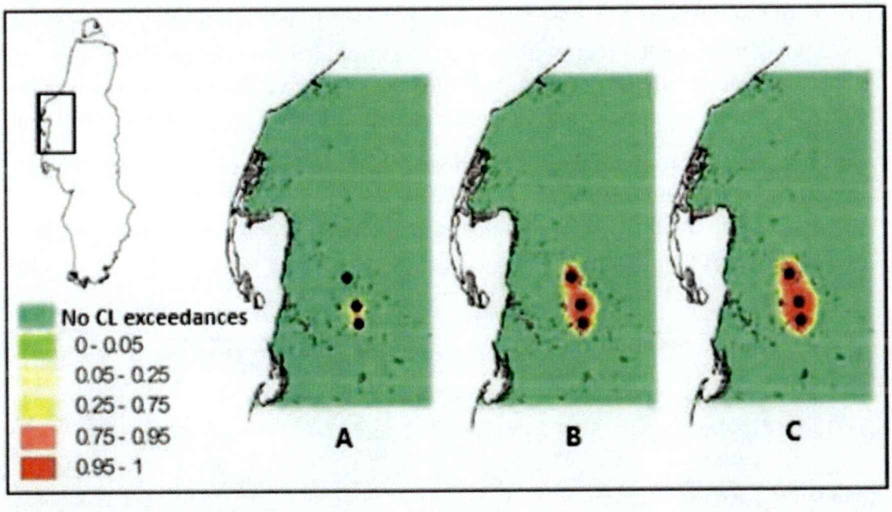

Fig. 3.39 The probability of exceeding the values of CL of nutrient nitrogen (%) during the phased introduction of objects: (**A**) start of operation, (**B**) intermediate stage, (**C**) maximum NO_x emission level

calculated for the most sensitive to nitrogen pollution oligotrophic species – lichens and mosses. High risks of eutrophication are predicted for ecosystems near BGCF after commissioning of booster compressor stations.

At the maximum level of emissions, the excess of $CL(N)_{nut}$ is significant for almost the entire 30-kilometer impact zone, which will be associated with a change in the structure of the vegetation cover of this territory due to the death of lichen species and mosses, as well as the disappearance of some oligotrophic vascular species (Bashkin, 2017a). The most likely effect of eutrophication of ecosystems in this area will be an increase in the number of sedges and cereals and an increase in the overall productivity of phytocenoses, which may lead to a change in the thermal characteristics of the soil and vegetation layer. This will be accompanied by an increase in the risks of loss of stability of foundations and pile foundations of many engineering structures. Another geoecological effect of increasing the atmospheric supply of technogenic nitrogen compounds in the zone of maximum impact may be an increase in the acidity of soil-ground and surface waters. Taking into account the fact that most of the atmospheric nitrogen deposition in this area is deposited in the snow cover, it is possible to predict the removal of part of the "excess" nitrogen with meltwater from terrestrial ecosystems to local reservoirs. Thus, the "actual" nitrogen loads resulting from the exposure of ecosystems to BGCF emissions will amount approximately to about 50% of the model values of precipitation.

When modeling precipitation, we did not take into account the (relatively small) part of nitrogen oxide emissions that is included in the processes of global dispersion in the atmosphere, which will increase the overall regional background of nitrogen precipitation in Yamal. Taking into account the data obtained, the increase in atmospheric nitrogen supply from the contribution of BGCF emission sources on a regional scale should not exceed 1–1.5 kg N/ha per year, which will not lead to a significant increase in environmental risks with respect to eutrophication effects (Fig. 3.40).

The dynamics of the productivity of the ecosystems of the impact zone during the phased commissioning of the integrated gas treatment plants will be as follows:

- Decrease in the productivity of biocenoses at the initial stages as the result of the death of lichens and mosses
- Increase in productivity at the final stages as the "response" of cereal species to additional nitrogen intake

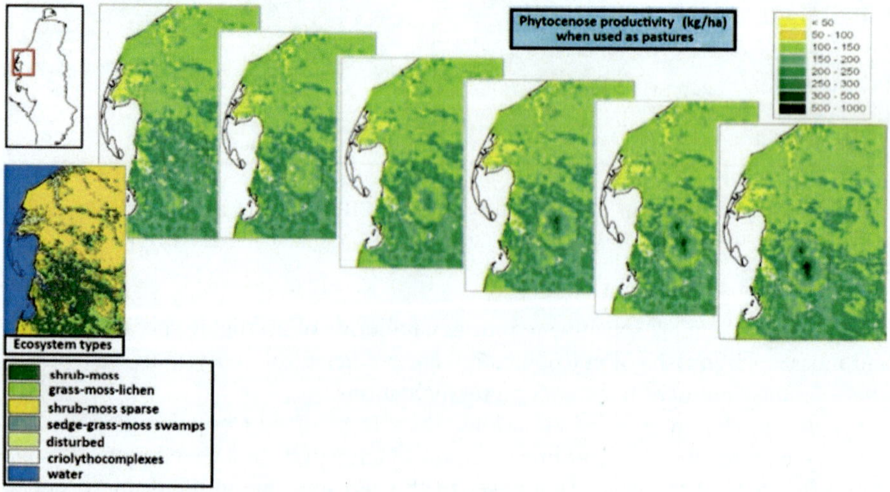

Fig. 3.40 Influence of NO_x emissions and atmospheric nitrogen deposition on the productivity of pasture lands in the BGCF impact zone

3.4 Assessment of Critical Loads in the Arctic Coastal Impact Zones of Main Gas Pipelines

Currently, there is an increase in the volume of natural gas transportation from the fields of the Far North of Russia to the central part of the country and a number of European countries, in particular, through the Nord Stream-1 and Nord Stream-2 gas pipelines. This is accompanied by an increase in the number of main gas pipeline lines (MP) and gas compressor stations (GCS), and as a result, an increase in potential geoecological risks. To the greatest extent, these processes will be observed in tundra, forest-tundra, and north-taiga ecosystems located in the northern parts of these gas pipelines. The variety of natural conditions of these territories and the conditions of their economic development determine the differentiated nature of ecosystem responses to the effects of GCS emissions and equipment used in the construction and operation of gas pipelines, which must be taken into account when analyzing potential environmental and geoecological risks in order to reduce and manage them. It should be borne in mind that the operation of compressor stations providing gas pumping through main gas pipelines involves the combustion of 0.0033 m^3 of gas per cubic meter of transported fuel with the release of nitrogen oxides NO_x (NO, NO_2) at discrete points of the pipeline route (Mingulov et al., 2005; Samsonov et al. 2007a, b). Taking into account the planned scale of natural gas transportation, the level of nitrogen oxide emissions only because of the planned work of GCS will amount to thousands of tons per year.

The degree of severity of geoecological risks for tundra and forest vegetation and soil biota, their structure and biodiversity in the area of exposure to GCS emissions will depend on many factors. Coniferous and small-leaved forests of the ecosystems under consideration, where main gas pipelines have already been laid, are usually deficient in nitrogen, which, along with harsh climatic conditions, determine their reduced productivity, for example, compared with similar Western European forests. Thus, among the potential effects of increased emissions and precipitation of nitrogen compounds in the areas of gas pipeline routes, an increase in forest productivity is possible, which from the point of view of their commercial use and carbon runoff can be considered as a positive effect. Along with the unfavorable disturbances of the biogeochemical nitrogen cycle discussed above, such complexity of geoecological risks associated with gas transportation through main gas pipelines complicates risk assessment and management procedures. These determine the increased importance of scientific substantiation of the parameters of permissible nitrogen inputs into ecosystems, taking into account the multifunctional nature of its compounds and environmental priorities of specific regions.

Similar conclusions can be made for tundra ecosystems; however, an increase in their productivity will occur due to the change of oligotrophic mosses and lichens to mesotrophic and eutrophic cereals. This will lead the destruction of the sod layer and the thawing of the upper permafrost layer. In these conditions, the importance of assessing and managing geoecological risks sharply increases, since such soil thawing is accompanied by a violation of the mechanical stability and integrity of gas pipeline pipes.

At the same time, it should also be taken into account that most of the natural indicators of the ecosystems under consideration are characterized by variability of possible values within certain intervals of values. It is associated with spatial heterogeneity of natural-territorial complexes at the macro and meso levels, species and physical characteristics of living organisms, as well as with the daily, seasonal, and interannual dynamics of the functioning of most components of the biosphere. Moreover, the parameters of the biogeochemical balance of elements in ecosystems also depend on a large number of biotic and abiotic factors. Taking into account the general concepts of geochemical ecology and ecological epidemiology (Kovalsky, 1974; Bashkin, 2006a, b), violations of the structure and/or functions of biological objects under external influences manifest themselves after exceeding a certain threshold or critical level of recipient exposure (Fig. 3.41).

Naturally, in complex biological systems (including ecosystems) characterized by a variety of components and the already mentioned variability of indicators in time and space, determining the magnitude of the critical level of impact as a reliable point value is very difficult, especially when it comes to estimates in a relatively long term (25–50 years). More correct parameters of permissible impacts – critical loads (CL) – can be obtained using probabilistic calculation methods that allow determining the most reliable range of possible CL values.

On the example of impact ecosystems in the zone of the Yamal–West main gas pipeline, the traditional method of calculating the values of critical loads in relation to eutrophying and acidifying effects is used to carry out probabilistic estimates of CL. For the input parameters of the mass balance equations and equations characterizing the intensity of migration flows of elements in ecosystems, the ranges of possible values are determined from the literature and cartographic data, which are included in the calculations using the Monte Carlo method. The 1000 model "runs" were carried out for each ecosystem, which gave a probabilistic distribution of

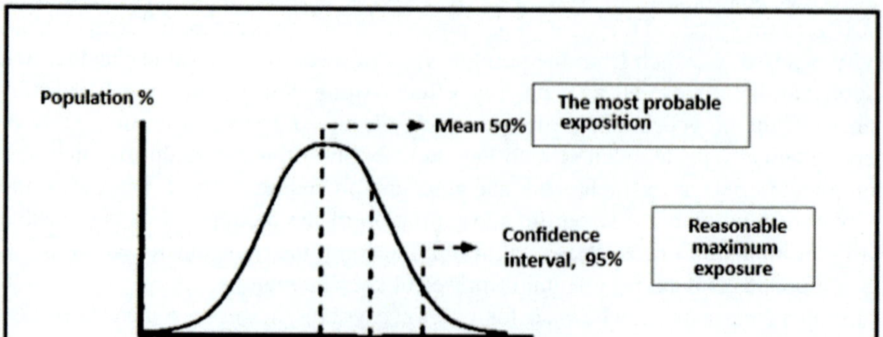

Fig. 3.41 Ecological principles of justification of acceptable levels of anthropogenic impacts of pollutants on ecosystems

possible CL values with a different combination of factors determining the resulting value. With all the randomness of the samples obtained, this ensured a high repeatability of the results.

The spatial resolution (detail) of the input information and the obtained values of CL corresponds to a spatial allocation of 1 × 1 km. When analyzing the obtained results of probabilistic estimates of the values of CL, 50, 75, and 95% levels of values corresponding to each "ecosystem" are considered. A corresponding GIS project has been prepared.

3.4.1 *Critical Loads of Acidity*

The distribution of CL values corresponding to 50, 75, and 95% levels of values is shown in Fig. 3.42.

According to the potential of neutralization of the acid component of atmospheric precipitation, the entire territory in the impact zone of the main gas pipeline can be divided into two parts: northern (tundra) and more southern (forest). Within the forest zone, the minimum values of critical loads of acidity are characteristic of swampy ecosystems. This differentiation is due to a whole complex of factors related to temperature conditions, moisture regime, soil texture, and vegetation types. According to the calculated values, the permissible level of acid deposition for tundra and swampy grassy ecosystems averages 100–200 g-eq./ha per year, whereas in forest ecosystems, the range of values obtained varies in the range from 300 to 700 g-eq./ha per year. This can be seen in Fig. 3.43 for GCS "Pechora," located in the northern part, and GCS "Ukhta," located to the south.

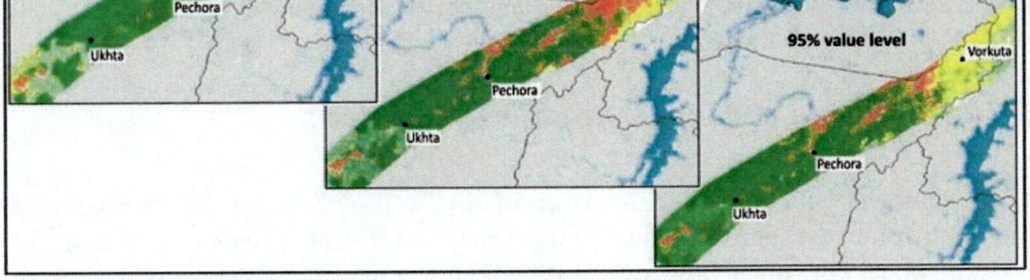

Fig. 3.42 Results of probabilistic calculations of CL of acidifying compounds (g-eq./ha per year). The detail of calculations is 1 × 1 km

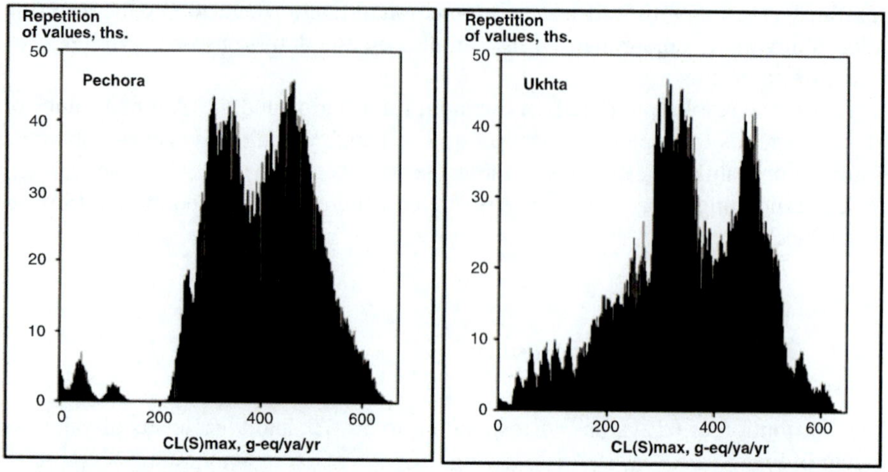

Fig. 3.43 Probabilistic assessment of ecosystem resilience to technogenic NO_x emissions in the Yamal-West GTS impact zone: acidification effects

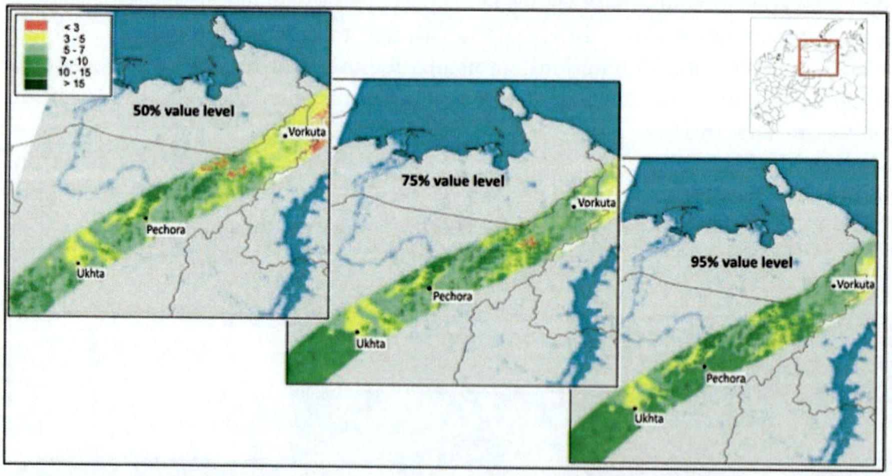

Fig. 3.44 Permissible parameters of precipitation of eutrophying nitrogen compounds for ecosystems in the impact zone of the Yamal-West GTS impact zone. Results of probabilistic calculations of CL of nutrient nitrogen (g-eq. / ha per year). The detail of calculations is 1 × 1 km

3.4.2 Critical Loads on Eutrophication Effects

The calculated value of critical loads in relation to the effects of the imbalance of nutrients for ecosystems in the tundra zone is 150–350 g-eq./ha per year (or 2–5 kg N/ha). For forest ecosystems, the allowable nitrogen intake is estimated at 350–700 g-eq./ha per year (or 5–10 kg N/ha). The obtained CL values (Fig. 3.44) are at the background level or slightly higher for these regions.

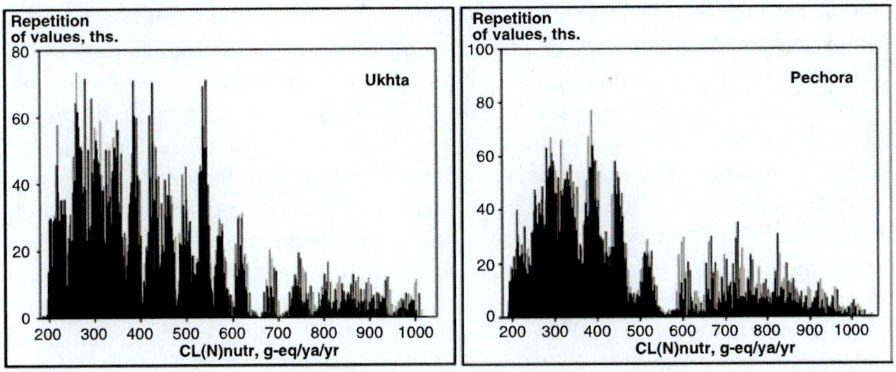

Fig. 3.45 Probabilistic assessment of ecosystem resilience to technogenic NO_x emissions in the Yamal-West GTS impact zone: eutrophication effects

The total distribution of probable values of CL of nutrient nitrogen for terrestrial ecosystems of two sites of Pechora and Ukhta GCS (the assessment zone is 50 × 50 km) is shown in Fig. 3.45.

The obtained results of the assessment of CL values allow us to rank ecosystems in the zone of impact of main gas pipelines according to their degree of resistance to atmospheric nitrogen deposition. This can be used in managing environmental and geoecological risks due to the planned increase in gas transportation volumes from the fields of the Far North to consumers in central Russia and European countries.

3.5 Conclusions

Thus, the use of methods for calculating geoecological risks and for assessing the probability of their excess in the impact zones of oil and gas projects in the Arctic region allowed us to draw a number of conclusions.

The analysis of data and methodological approaches to the assessment of geoecological risks in connection with the anthropogenic impact of LNG production and natural gas production in the shelf zone of the Arctic seas revealed the differentiated nature of the stability of various aquatic ecosystems in relation to atmospheric emissions of nitrogen oxides.

The greatest geoecological risks (in terms of the degree of manifestation of negative changes in the structure and functioning of ecosystems) are associated with the impact of the LNG plant on freshwater (and terrestrial) ecosystems of the northern coast of the Kola Peninsula. For coastal areas of the Barents Sea, the most likely geoecological risks are associated with the eutrophication of these ecosystems under the maximum exposure scenario, which is associated with the already existing high level of anthropogenic impact in the region. Taking into account the approaches used, relatively low levels of eutrophication risks for benthic communities of marine ecosystems can be predicted in the area of the Shtokman field in the water area within 5–10 km of the zone from the emission source.

The impact of existing gas production and transportation projects on the probability of geoecological risks was carried out in the coastal Arctic regions: the Taz Peninsula and the Yamal Peninsula. It is shown that these risks largely depend on the intensity of the work of the UCP. The probability of geoecological risks is higher in impact zones adjacent to point sources of nitrogen oxide emissions. In some cases, due to the effects of vegetation change, there is a high probability of risks of soil thawing, which can lead to a violation of the stability of piles of gas pipelines and foundations of engineering structures and road facilities. This, in turn, requires the reclamation of such ecosystems (Bashkin & Galiulin, 2019).

Chapter 4
Environmental Rating as an Indicator of Geoecological Risk Management of Russian Oil and Gas Companies in the Arctic

Abstract This chapter focuses on the issues of environmental rating as an indicator of geoecological risk (GER) management of Russian oil and gas companies in the Arctic. The authors discuss the need of using environmental rating for oil and gas industry in Russia. Primarily, specialized attention is given to the issues of rating results of Environmental Responsibility of Oil and Gas companies in Russia that was first held in 2014 by the cooperative initiative of CREON Group and WWF Russia along with National Rating Agency. The authors have selected and analyzed from all Russian oil and gas companies only those who are operating in the Arctic region.

The rating's results show the "top three" companies among Russian oil and gas companies operating in the Arctic – Zarubezhneft, Lukoil, and Sakhalin Energy. There are also other companies being rated. For instance, Zarubezhneft and LUKOIL share the best position in terms of environmental management (the 1st and the 2nd position). The best companies in "environmental impact" sector are Gazprom, Zarubezhneft, and LUKOIL, whereas the leaders in "disclosure/transparency" sectors are Sakhalin Energy (Sakhalin-2) and Zarubezhneft (the 1st and the 2nd position).

In whole, the environmental rating of the Russian oil and gas companies, operating in the Arctic, may create a new mechanism of identification of their CSR level. Particularly, analysis of a number of criteria such as environmental management, environmental impact, and information disclosure or transparency helps correlate the process and results of economic activity of oil and gas companies in the Arctic with the requirements and expectations of society. Furthermore, the desire to increase these indicators contributes to investment attractiveness of oil and gas companies and testify about their environmental responsibility. It helps to inform foreign investors, and resulting in the rise of FDI in Russia as well as decline the geoecological risk (GER) in conditions of Arctic's exploitation.

Keywords Environmental rating · Oil and gas industry · Geoecological risk · Investment attractiveness · The Arctic · CSR

V. N. Bashkin, O. P. Trubitsina, *Geoecological and Geopolitical Risks for the Oil and Gas Industry in the Arctic*, Environmental Pollution 29,
https://doi.org/10.1007/978-3-030-95910-4_4

4.1 Introduction

At present, the global crisis has led to a reduction in forecasts of economic growth and energy consumption in the world. Moreover, the current predictions are ambiguous and are dependent on a combination of many factors, primarily, geopolitical and environmental. It is critically important to consider these issues for energy security, even if the world shifts to an electrified, renewables-rich energy system.

According to the Global and Russian Energy Outlook (2014), this connects with energy transitions and affects to oil and gas demand: therefore, supplies become more concentrated in a smaller group of resource-rich countries even if their economies simultaneously come under strain due to lower export revenues. In the long term, fossil fuels will continue to dominate with a slow increase in the share of non-carbon energy resources. Particularly, share of oil and gas in global primary energy consumption will remain virtually unchanged (53.6% in 2010 and 51.4% by 2040). It should be noted that the baseline scenario is that the global gas consumption is projected to grow to 5.3 trillion cubic meters by 2040, which will be more than 60% higher than that in 2010. First, such a rapid increase in gas demand in all regions will be owing to the development of gas generation for reasons of increasing electrification and, therefore, growth in electricity consumption. At the same time, the demand for gas consumers in industry will also grow rapidly in developing countries. The environmental advantages of gas will support its role in individual markets as well.

Regardless to predicted cyclical downturn oil prices that has started in 2020 (Commodity cycles, 2013), at present one see the relevant growth in 2022, and accordingly, a collapse in oil prices to $24 per barrel by 2050 is possible only if the implementation of the most radical decarbonization scenario worldwide. In all other cases, they will remain quite high, as calculated by the IEA. In the short term, it has been predicted that the level of world energy demand in 2021 has recovered compared to that in the previous year, which means growth by 4%. In a result, world energy demand is returning to pre-pandemic levels now, which means a rise in the economic activities – i.e., a rise in the demand for all fuels and technologies (World Energy Outlook, 2021) – as the countries are gradually lifting their lockdowns.

Thus, in view of energy demand, the oil and gas industries are being challenged to improve production efficiencies and to maintain a growth trajectory in its crude oil and natural gas resources. For this purpose, it is important to be able to address a number of operational challenges so that future demand is met with adequate supply (Improving safety, 2014). Human and environment safety and health protection remain the dominant priority for oil and gas industries (Bigliani, 2013).

Russia is one of the major oil and gas producing countries in the world according to the International Energy Agency (Wernstedt, 2002). In essence, the Arctic is a field of new opportunities along with the geoecological and geopolitical challenges for oil and gas companies. It is expected that the region will be playing a vital role in meeting Russia's energy supply over the next several decades. That is why it is important to outline the main problems and find ways for improvement in the oil and gas sector. The companies that operate in the Arctic should be analyzed in the sphere of ecological awareness of the consequences that might be faced when oil and gas are exploited there.

The activity of the oil and gas industry in the Arctic Zone of the Russian Federation (AZRF) has increased in the past several years and the GER has been increasing accordingly. The Russian Arctic is considered a relatively pristine area compared to many other parts of Russia. However, every environmental negative effects caused by the industries in the AZRF can lead to the formation of "hot spots" and "impact areas" characterized by a high level of chemical contamination of the environment and transformation of the natural geochemical background (Diagnostic analysis, 2011; Trubitsina, 2016a, b). For example, there is always a variety of potential GER and negative impact on the environment due to oil and gas exploration. At the same time, consequences depend upon the stage of the process, the size and complexity of the project, the nature and sensitivity of the surrounding environment, and the effectiveness of planning, pollution prevention, mitigation, and control techniques (Environmental Management, 1997). Similarly, the effects of marine pipelines in the Arctic on the biota are distinguished into two groups based on their effects on natural systems: (1) effects due to mechanical operations (increased water turbidity due to high concentrations of a suspended bottom sludge at excavation of trenches in shallow areas) and (2) hydrocarbon effects (Trubitsina, 2016a, b). Taking everything into account, the GER for oils and gas objects, operating in the Arctic indicators, is defined as risks that occur in the environmental arctic conditions in the "industry– environment" system, as related to the mutual impacts of industrial facilities on the environment and of the environment on industrial facilities (Bashkin, 2014; Trubitsina & Bashkin, 2016).

At the same time, Russian oil and gas industry also is faced with barriers to foreign direct investment (FDI) due to the uncertainty of the existing legal regime of Arctic, particularly, geopolitics challenges and the trustfulness of technologies that comply with international environmental standards. For example, underinvestment may lead to oil spills and pipeline leaks that influence the increase in GER. This chapter is devoted to certain methods used to assess the outside of investment appeal of Russian oil and gas companies operating in the Arctic, particularly the use of existing assessment results provided by credit rating agencies, leading journals, etc. Accordingly, this chapter focuses on issues of environmental rating (ER) as an indicator of GER management of Russian oil and gas companies in the Arctic. ER is also discussed in view of a factor of improving foreign investment attractiveness of companies by raising environmental awareness of business such as by focusing on activities connected to striving toward the reduction of GER and sustainable development of the Arctic region.

4.2 What Are the Motivations of Oil and Gas Companies Operating in the Russian Arctic to Develop Corporate Social Responsibility Strategies?

Most experts believe that merging of corporate social and environmental responsibility (CSR) principles and practices of large businesses in the Russian Arctic will be a solution for many problems associated with these companies' activities (Sergunin, 2021). CSR is a corporate approach that includes activities that benefit social well-being and the environment. What are the motivations of oil and gas companies operating in the Russian Arctic to develop corporate social responsibility strategies?

Let us consider the main ones:

1. Economic sociology understands the CSR as a form of social partnership, that is, the responsibility the companies have toward the expectations of a society. Companies operating in the Russian Arctic and receiving substantial benefits from it bear obligations not only to their employees but also to other people living in this territory including indigenous peoples and are responsible for the environmental safety. In the event, if this partnership does not develop, there is a ground for conflict between companies, on the one hand, and local authorities and society, on the other. Therefore, such a conflict is not a benefit for both sides. There is similar theory that considers CSR as a method of benefits sharing. According to this approach, companies that benefit from their economic activities in a certain territory (in our case, in the Russian Arctic) cannot appropriate this benefit only to themselves. If they desire to continue their activities in this region on a stable basis, they must share their benefits with the local population and take care of improving their living conditions, including social and environmental situation.
2. In sociology and political science, the theory of corporations as "good citizens" of society (Matten & Crane, 2005) has also spread. According to this theory, each social group performs its own function in society. Along with it, every social actor including business must faithfully perform its functions in order for society to develop normally. Therefore, companies operating in the Russian Arctic are obliged not only to develop the economy of the region but also to take care of the welfare of the local population and the preservation of the natural wealth of the Russian North.

Modern versions of this theory believe that CSR logically follows from corporate ethics, the role of which is increasing in the current conditions. Supporters of this corporate ethics explains CSR as follows: "modern ("civilized") business is characterized not only by the pursuit of profit, but also by the understanding that it bears certain moral obligations to other members of society" (Freeman, 2004).

It should be noted that the theory of "good citizen" has become popular especially in the social sciences due to the fact that in cases of global crisis, the state as a source and guarantor of social benefits (the concept of the "welfare state") transfers part of the authority to someone to take over part of the social functions of the state and ensure the progressive development of society (Sergunin, 2021; Kumpula et al., 2011). In accordance with this, oil and gas business should strive to achieve high international social, environmental, and ethical standards in this area, especially if they are active in the Arctic.

The assumption of the role of "good citizens" by companies contributes to the elimination of negative phenomena such as corruption, fictitious tenders, "black" and "gray" economy, opacity of corporate and government decisions, lack of participation of civil society institutions in the development of these decisions, and ignoring of the social and environmental aspects of industrial and infrastructure projects in the Russian Arctic (Sergunin, 2021).

3. At present a set of theories explaining CSR by business or managerial logic has become widespread. So, some researchers suggest that not only are the management, shareholders, and employees of a particular company but also the social environment in which it operates, i.e., local and regional authorities, civil society institutions, especially vulnerable groups of the population, are considered to be a direct and important stakeholder in corporate decision-making (Stammler & Peskov, 2008). If the company takes into account the interests of these informal "shareholders," the quality of its decisions will increase, the likelihood of conflicts with the local population and authorities will decrease, and, as a result, the conditions for the continuation of business will become more favorable.

It is interesting to note that CSR is considered part of the corporate risk management strategy by similar theory (Sergunin, 2021). The more channels a company has for interacting with local authorities and communities, the less risks are associated with the uncertainty of the social and institutional environment and the more the business feels safe in this environment. In accordance with this theory, a company that cares about the local population and the environment is more attractive as a partner for authorities at all levels and other companies. Therefore, such company has a better chance of success in obtaining lucrative contracts or promoting its products and services on the AZRF market.

4. Finally, on the grounds of sustainable development concept that was approved at the official level by both international organizations (primarily the UN and its specialized agencies) and national governments, including Russia, CSR is understood as the harmonization of economic, social, and environmental responsibility of all social actors equally in these three areas. From the point of view of this theory, CSR is seen as a natural duty of business to ensure the sustainable development of the AZRF (Konyshev et al., 2017). By analogy with entrepreneurs, other participants in Arctic activities (individuals, social groups, institutions, organizations) should also contribute to the achievement of this goal – of course, to the best of their abilities and capabilities.

Social responsibility has different levels and is the combinations of companies' requirements and society's expectations, along with the profitability/disadvantage of oil and gas enterprises due to their social activity in the Arctic. The higher the level of social responsibility of companies, the more they have voluntarily assumed obligations. In accordance with this, it is possible to imagine social responsibility like a pyramid which has three levels (Fig. 4.1).

The first level is the lower, basic, mandatory component of social responsibility. It includes compliance with laws tax legislation, labor and civil codes, environmental norms, all related to the obligations established for their activities in the Arctic initially. The second and third levels refer to voluntary components of social responsibility. These two stages differ in the motives for implementing socially responsible behavior. The second level is the realization of socially responsible behavior for the sake of economic benefit mostly for the following purposes:

- To create a "green image" of the organization for its consumers and employees
- To increase investment attractiveness
- To create favorable relations with the government and society

The third level covers socially responsible behaviors that are not aimed at obtaining economic benefits. This is a higher level, and it considers the position and role of the organization in society. Such actions can bring economic profit, but that is not their goal. In this case, oil and gas companies involve in charity aimed at helping people inhabiting the Arctic, including indigenous peoples, as well as implementing environmental measures to ensure the preservation of the environment.

Thus, interactions within "industry-environment-society" system are based on the combined use of imperative and dispositive methods of legal regulation. Mandatory regulatory requirements mainly reflect the average level of the market and do not allow companies to minimize the costs of compliance with

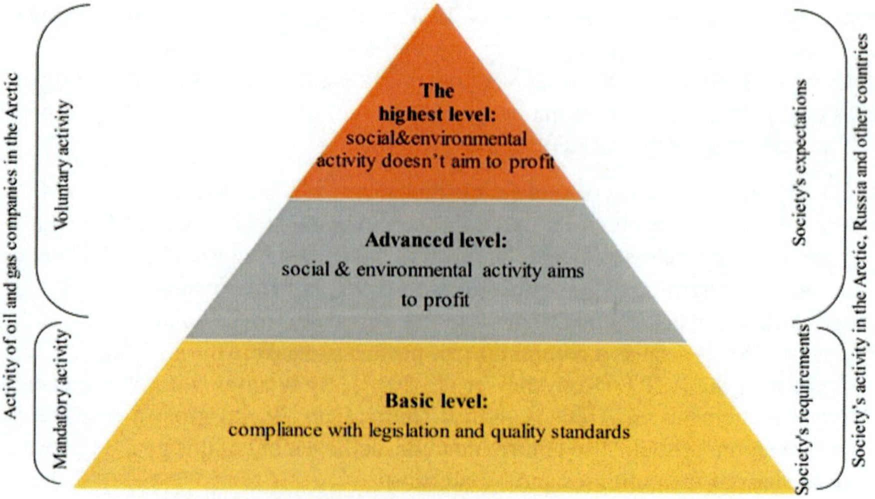

Fig. 4.1 Levels of social responsibility of oil and gas companies operating in the Arctic based on (Bratyushchenko & Seliverstov, 2007)

environmental responsibility standards. To strengthen their competitive advantages in environmentally sensitive markets, companies choose to adopt voluntary environmental standards (Sharkey & Bromley, 2015). For instance, even a few decades ago, the environmental statistics in oil and gas industries was not open in various countries, including Russia (Douma, 2010) and at present due to developing voluntary environmental programs, their comparative perspectives are looking much more transparent.

4.3 Environmental Responsibility Rating of Russian Oil and Gas Companies Operating in the Arctic

Voluntary submission of reports within the framework of raising environmental awareness of the oil and gas industry and meeting the society's expectations are essential for increasing the level of social responsibility and environmental management in companies operating in the Arctic. At the same time, filling the deficit of objective information makes it possible to use environmental criteria more effectively in the markets of investments, equipment, goods, and services, especially during the recovery period after the COVID-19 crisis.

It should be noted that the main risks among others for oil and gas industry enterprises operating in the Arctic region are geoecological risks (GER) and geopolitical ones (Trubitsina & Bashkin, 2017a, 2017b, 2018, 2019a, 2019b, 2019c, 2019d, 2020). Thus, as a part of the recovery of the industry after the crisis, it is necessary to organize a management system of GER, as well as GPR at oil and gas industry facilities, formulate effective anti-crisis strategies.

The environmental rating, as a neutral tool, allows assessing the level of CSR and, consequently, environmental risk management, and investment attractiveness of oil and gas companies by determining their transparency and environmental responsibility. It is well known that the key indicators of investment decisions in the energy sector are responsible for the governance of environment, society, and corporate.

4.3.1 Rating Objectives

The environmental responsibility rating (ERR) of Russian oil and gas companies has been conducted jointly by the CREON Group and the WWF of Russia along with the National Rating Agency since 2014.

Only those Russian oil and gas companies that operate in the Arctic region were selected and analyzed according to ERR for the period 2014–2020. ERR is considered a tool for assessing the level of CSR, in particular the management of GER so as to ensure environmental safety in the Arctic region. Overall, the analysis aims to identify the general and differentiated dynamics of greening in the Arctic oil and gas sector.

4.3.2 The Rating Method

In order to compare the dynamics of the rating results, the period 2014–2020 was analyzed. Only seven companies with published statistical indicators were considered based on two key criteria: (1) the activities of the companies in the Arctic region should be both on the Arctic shelf and in the Russian Arctic and (2) the lower limit of the production volume should be 1.5 million tons (2014–2019), oil equivalent (oil and gas condensate) be 2 million tons (2020), the volume of oil transportation be 30 million tons per year, and the volume of refining of oil, gas condensate, and petroleum products be 8 million tons per year.

It is important to note that the number of Russian oil and gas companies selected for the rating varies over time. For instance, the gas company “Arktikgaz” was included in the rating in 2015 and excluded in 2019 because it collaborated with “Novatek” in 2018. The oil company “Bashneft” was included in the rating in 2014 but not in 2016 due to its collaboration with “Rosneft”. In addition, a number of companies own other companies, for example, Gazprom-Sakhalin Energy, Gazprom Neft. Furthermore, all companies are considered independent, since each retains its own internal policy as well as a freely formulated CSR policy (Shvarts et al., 2015).

The companies’ activities were categorized into three sections according to Shvarts et al. (2016) for the purpose of analysis. The first one is environmental management. It is important to note that most criteria followed by companies of this section are more rigid compared to the Russian legislation on environmental protection. They correspond to the best global standards and practices in the oil and gas business.

The second one is environmental impact, which evaluates the scale of impact of oil and gas companies on the environment, particularly the level of damage on air, water, and land resources during project implementation. In most cases, the criteria are based on components of official statistics on environmental protection.

The third one is disclosure/transparency. This section evaluates the extent of the readiness of the companies to disclose information about the environmental impact of their industrial activities (Knizhnikov et al., 2015). The rating’s ecological criteria are based on Environmental Standards for Operations (2005), a document developed by a coalition of non-governmental environmental organizations as a common environmental public standard for improving oil and gas companies’ environmental policies, procedures, and practices.

4.3.3 Analysis of Rating Results

Rating results of the first and final years of the study period for oil and gas companies, operating in the Arctic, are compared in Table 4.1.

The most visible trend of the rating held in 2020 comparing to 2014 is the positive dynamics of the final rating points for all analyzed companies, even the number of rated companies has decreased (Table 4.1). Such improvements in the rating might be due to the collaboration of companies and summation of their rated indicators. The three top-rated companies in 2020 among other companies operating in the Arctic were Zarubezhneft, Lukoil, and Sakhalin Energy, while the best ones in 2014 were Sakhalin Energy, Gazprom, and Rosneft (Trubitsina & Bashkin 2017a, b). Zarubezhneft showed the biggest difference in the final rating indicator for the years 2020 and 2014. It should be noted that improvements observed in the rating are mainly due to disclosure of additional information by the companies on environmental management and environmental impact. For instance, Zarubezhneft has not only shown improvement in disclosing their annual non-financial information, but also they have made public on the Internet a large volume of additional information regarding their subsidiaries.

Table 4.2 shows the 2020 rating results according to the three sections of oil and gas companies operating in the Arctic

It is interesting to note that the leading companies in all three sectors are different. Zarubezhneft and LUKOIL (Sections 1 and 2) occupy the best positions in environmental management sector. Gazprom, Zarubezhneft, and LUKOIL are the leaders in "environmental impact" sector. Gazprom is one leading state-owned gas producer, which accounts for approximately 70% of total gas production in Russia and owns approximately 11 large oil extracting companies in the country (Shvarts et al., 2015). It has been maintaining its position since 2014 and currently it is the leader of the sector "environmental impact". Section 3 demonstrates the level of corporate transparency with respect to environmental issues, which is largely

Table 4.1 Comparison of the rating results for the years 2014 and 2020 for the Russian oil and gas companies operating in the Arctic

Company	Final 2020 position/ final 2020 rating point	Point change as compared to 2014 result	Final 2014 position/ final 2014 rating point
Gazprom[a,b]	4/1.5210	▲ 0.1665	2/13545
Sakhalin Energy (Sakhalin-2) [a,b]	3/1.7304	▲ 0.2051	1/15253
Zarubezhneft[b]	1/1.8232	▲ 0.7678	4/10556
LUKOIL[b]	2/1.7521	▲ 0.7578	5/0.9943
Rosneft[a,b]	5/1.5025	▲ 0.4165	3/10860
Gazprom Neft[a,b]	6/1.4690	▲ 0.5973	6/0,8717
NOVATEK[b]	7/1.3130	▲ 0.5925	7/0,7205

[a]Companies operating in Arctic offshore
[b]Companies operating in the AZRF

Table 4.2 Rating results for 2020 according to three sections of oil and gas companies

Company	Section 1 Environmental management		Section 2 Environmental impact		Section 3 Disclosure/ transparency	
	Section 1 position	Rating point	Section 2 position	Rating point	Section 3 position	Rating point
Gazprom	6–7	1.3750	1	1.7436	6	1.4444
Sakhalin Energy (Sakhalin-2)	3–4	1.8750	4	1.5385	1–2	1.7778
Zarubezhneft	1–2	2.0000	2	1.6923	1–2	1.7778
LUKOIL	1–2	2.0000	3	1.5897	3–4	1.6667
Rosneft	3–4	1.8750	7	1.0769	5	1.5556
Gazprom Neft	5	1.6250	6	1.1154	3–4	1.6667
NOVATEK	6–7	1.3750	5	1.2308	7	1.3333

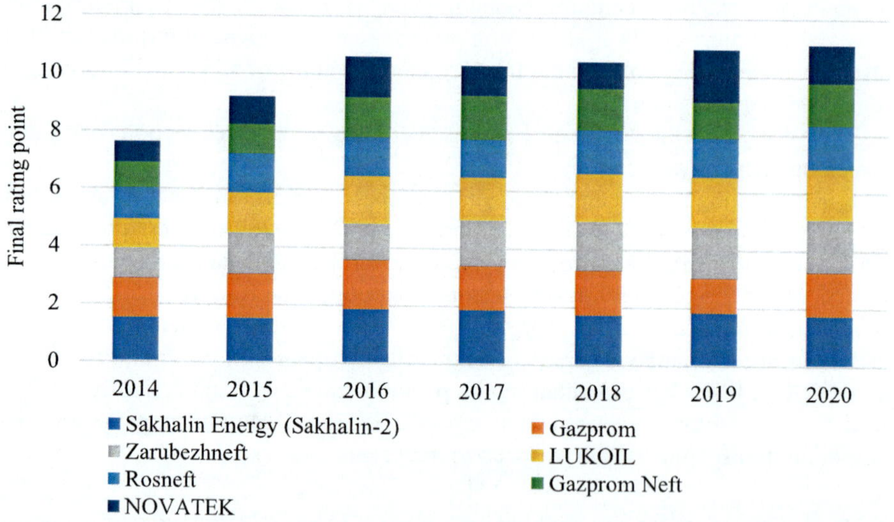

Fig. 4.2 Dynamics of ERR results of oil and gas companies operating in the Arctic for 2014–2020

determined by the quality of non-financial reporting. There are three types of such reporting for Russian oil and gas companies: environmental reports, social reports, and sustainable development reports. The latter are the most popular form (RSPP, 2015; Shvarts et al., 2016). All the analyzed companies disclose environmental and social parameters of their activity in one way or another. The leaders in sector "disclosure/transparency" are Sakhalin Energy (Sakhalin-2) and Zarubezhneft (Sections 1 and 2).

Analysis of the ERR results conducted in 2014–2020 is shown in Fig. 4.2, which shows the dynamics of indicators based on three sections: (1) managerial, (2) operational, (3) and informational. In general, the total indicators of final ratings point are growing positively in dynamics. Regardless of the fact that the leaders change from

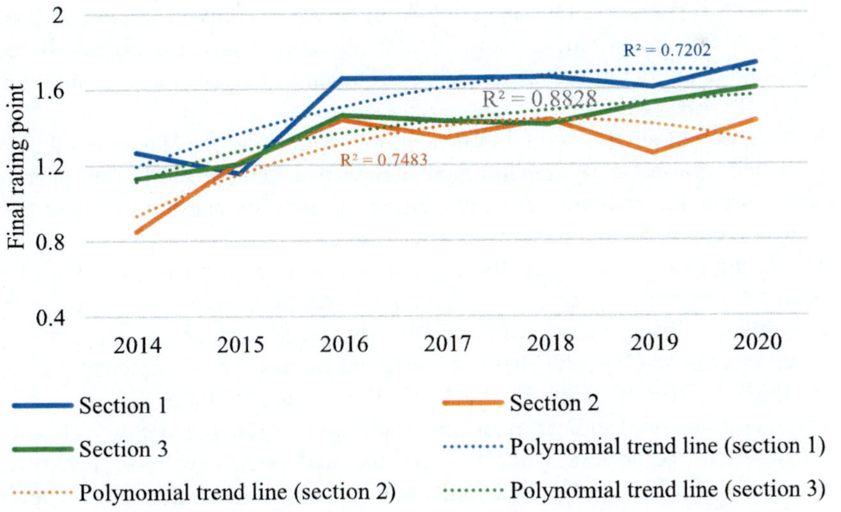

Fig. 4.3 Dynamics of the average annual indicator of the final rating points calculated for the rated oil and gas companies operating in the Arctic for 2014–2020

year to year, companies with consistently high indicators of ERR are identified, for example, Sakhalin Energy (Sakhalin-2), Zarubezhneft, and Gazprom.

The dynamics of the average annual indicator of the final rating points calculated for the rated oil and gas companies operating in the Arctic for 2014–2020 are shown in Fig. 4.3.

It becomes possible to trace the trends in the development of final ratings points (FRP) for individual sections. In the first section, the FRP fell significantly only in 2015, and after a sharp rise in 2016, it became stabilized until 2020. It should be noted that it was during this period (from 2016 to 2020) that FRP in the second section was unstable compared to the first. It is likely that in the first two years of economic activities in the licensed areas the rating oil and gas companies were more focused on compliance with the requirements reflecting the impact on the Arctic environment because of implementation of economic activities in the licensed areas of the oil and gas complex. In third section, there is an annual increase in the FRP, which indicates the willingness of companies to disclose information about the environmental impact during production activities.

4.4 Conclusion

At present, due to increase in energy demand, the oil and gas industries are being challenged to improve their production efficiency and to maintain a growth trajectory in their crude oil and natural gas resources. In this context, Arctic is an area of new opportunities for oil and gas companies. However, Russian oil and gas industry is faced with barriers to FDI, appearing from the uncertainty of the existing legal regime of Arctic, in particular, geopolitical challenges and the trustfulness of technologies that comply with international environmental standards and ensured minimization of GER.

Introduction of environmental ratings allows us to prioritize and assess the level of CSR and, consequently, environmental risk management, GER minimization, and investment attractiveness of oil and gas companies, by determining their transparency and environmental responsibility.

The rating results show that there are differences among Russian oil and gas companies operating in the Arctic. It depends on their levels of environmental responsibility and transparency. Large, publicly traded companies with a strategic focus on increasing CSR level have received top ratings. The companies that were rated "top three" for 2020 are Zarubezhneft, Lukoil, and Sakhalin Energy.

In general, the total indicators of final rating point grew positively in dynamics for 2014–2020. Regardless of the fact that the leaders change from year to year, companies with consistently high indicators of ERR are identified, for example, Sakhalin Energy (Sakhalin-2), Zarubezhneft, and Gazprom.

Different companies are included in the rating. Zarubezhneft and LUKOIL (Sections 1 and 2) share the best position in terms of environmental management The best companies in the "environmental impact" sector are Gazprom, Zarubezhneft, and LUKOIL, whereas Sakhalin Energy (Sakhalin-2) and Zarubezhneft (1–2 section position) are the leaders in "disclosure/ transparency" sector.

Comparison of the results of the ERR rating of oil and gas companies operating in the Arctic with the CSR level shows that all the above companies occupy the first basic level, which is an obligatory component of social responsibility. Will they rise to higher levels? The fact that companies voluntarily submit reports as part of raising environmental awareness and business responsibility indicates their inclusion in the second or third level of CSR. However, more in-depth research is required for their exact affiliation to an advanced or higher level. Nevertheless, it is important to note that transparency in the disclosure of environmentally significant information in the oil and gas sector is growing, which means that the companies are climbing to the top of the CSR pyramid.

Finally, it should be added that the theory of "good citizen" is more connected with current situation in Russia and world. Particularly, when there is a global crisis, the state as a source and guarantor of social benefits must give over part of the social functions of the state for ensuring the progressive development of society. In accordance with this, oil and gas business should strive to achieve high international social, environmental, and ethical standards in this area, essentially if they are active in the Arctic.

Chapter 5
Management of Geopolitical and Geoecological Risks in the Development of Natural Resources in the Arctic

Abstract This chapter shows the examples of geopolitical and geoecological risk management, affecting the three basic directions in the work of oil and gas companies, in particular, gas producing and gas transportation companies – technological, geoecological, and environmental as well as reveal the need to revise a number of provisions on the assessment of various risks of Arctic and subarctic oil and gas projects. Geoecological risk management on complex main gas pipelines is also demonstrated. Management of geopolitical and geoecological risks as a basis in the development of natural resources in the Arctic is also shown. Environmental risk insurance as a risk management mechanism is considered for the given projects. Taking into account the existing realities of the Arctic region, geopolitical and geoecological risks shown are among the priorities, the management of which will make the greatest contribution to the development of oil and gas projects.

Keywords Management of geopolitical and geoecological risks · Oil and gas arctic projects · Modeling · Insurance

5.1 Features of Analysis and Risk Management of Oil and Gas Projects in the Arctic Region

Risk management in oil and gas projects in the Arctic region requires to take into account the geopolitical, climatic, ethical, institutional, geo-ecological, technological, financial, and other aspects of project execution, which makes increased demands associated with the analysis of various nature risks and their relationship (Meshcherin et al., 2011; Kazak & Bashkin, 2015a, b).

V. N. Bashkin, O. P. Trubitsina, *Geoecological and Geopolitical Risks for the Oil and Gas Industry in the Arctic*, Environmental Pollution 29,
https://doi.org/10.1007/978-3-030-95910-4_5

On the other hand, oil and gas projects in this region are characterized by increased capital intensity, the duration of the investment cycle, and large number of stakeholders/as well as high volatility of energy prices. These features cause an increased level of risks of these Arctic projects, and their effective implementation is simply impossible without the use of the most modern and advanced risk management methods and tools, including geopolitical and geoecological risks (Bashkin & Kazak, 2015; Bashkin & Lesnych, 2015).

It should be emphasized once again that risk in this case is understood as an uncertain event or condition that, if it occurs, has both a positive and a negative impact on at least one of the goals of the project (Meshcherin et al., 2011; Trubitsina, 2016).

Management of all the mentioned risks in the conditions of the Arctic region should be carried out based on the following principles:

1. Creation of benefit. Risk management of Arctic oil and gas projects significantly increases the level of their manageability and increases the likelihood of achieving their goals, which, in turn, creates additional project costs.
2. The concept of acceptable risk. In the conditions of the Arctic region, it is impossible to avoid all risks when managing oil and gas projects. In this regard, it is necessary to determine what level of risk is acceptable for the project.
3. The coverage of the whole set of risks means that in order to reduce the uncertainty of achieving the project goals, it is necessary to take into account all areas of risk occurrence.
4. The use of opportunities implies that when managing the risks of Arctic oil and gas projects, it is necessary not only to reduce possible threats but also to strengthen the potential capabilities of these projects;
5. Consistency with the external environment. The external environment is one of the main sources of risks of Arctic oil and gas projects; therefore, the process of managing them should be built in accordance with the requirements put forward by the external environment.
6. Openness to stakeholders who determine the goals and success of Arctic oil and gas projects. Risk management should increase their awareness of the progress of the project in order to be able to make appropriate changes to it in a timely manner.
7. Adaptation to changes. Throughout the life of the project, changes can occur both in its internal and in its external environment. In order to achieve the project objectives, the risk management system must be able to adapt to the changes taking place.

The general algorithm for risk analysis of Arctic oil and gas projects consists of six processes: risk management planning, risk identification, qualitative and quantitative risk assessment, risk response planning, monitoring and risk management (Fig. 5.1).

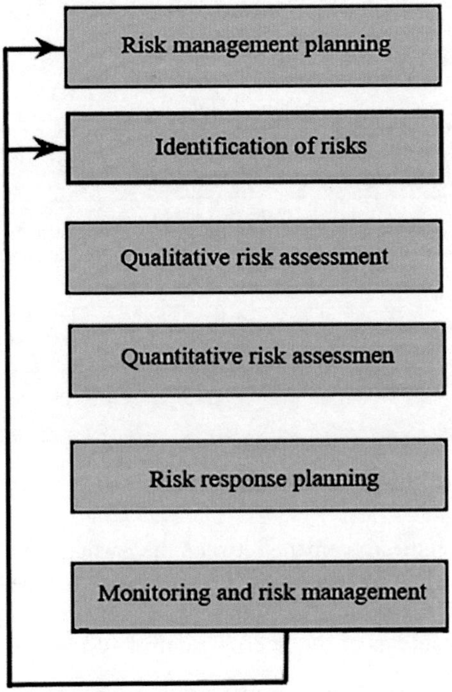

Fig. 5.1 General algorithm for risk analysis of Arctic oil and gas projects

5.1.1 *Risk Management Planning for Arctic Oil and Gas Projects*

By the nature of the occurrence, the following risks of Arctic oil and gas projects can be distinguished:

- Economic – representing the possibility of economic losses due to the accidental nature of the results of economic decisions or actions taken
- Geopolitical – characterized by the possibility of losses or a reduction in profits resulting from the implementation of state policy in the countries participating in Arctic oil and gas projects or their competitors
- Technological – risks caused by technical reasons
- Natural – representing the possibility of undesirable consequences from adverse and dangerous natural phenomena and processes
- Geoecological – issues arising in the "oil and gas industry– Arctic environment" system related to the interdependent impact of industrial facilities on the environment and the environment on industrial facilities
- Social – risks which are related to social rights and (or) social guarantees of people

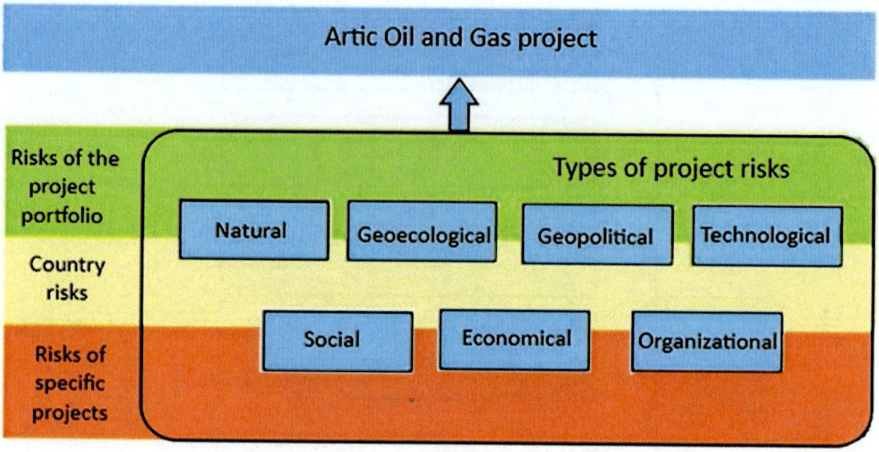

Fig. 5.2 Risk structure of Arctic oil and gas projects

- Legal – the risk that the agreement between the participants will be impossible to fulfill under the current legislation or the agreement will be executed improperly
- Organizational – risks associated with mistakes of the company's managers, its employees, and problems of the internal control system

The risks of the projects under consideration may arise at various levels; in accordance with this, they are distinguished (Fig. 5.2) as follows:

- Risks of portfolios of Arctic oil and gas projects
- Country risks (sub-portfolio level of projects)
- Risks of specific projects

A generalized risk management planning algorithm for Arctic oil and gas projects is shown in Fig. 5.3.

5.1.2 Identification of Risks of Arctic Oil and Gas Projects

During identification, the main risk factors that can affect the project are determined. The risk identification process is carried out at all stages of the project life cycle, because new risks may be discovered during its implementation.

To identify the risks of Arctic oil and gas projects, the following methods and tools can be proposed:

- Analysis of project documents, analysis of modern scientific literature on the assessment of the external environment of the project (geopolitical and geoecological risks), analysis of own data and other sources. Compliance of the project plans with its requirements and assumptions is an indicator of the possibility of the presence of risks in the project

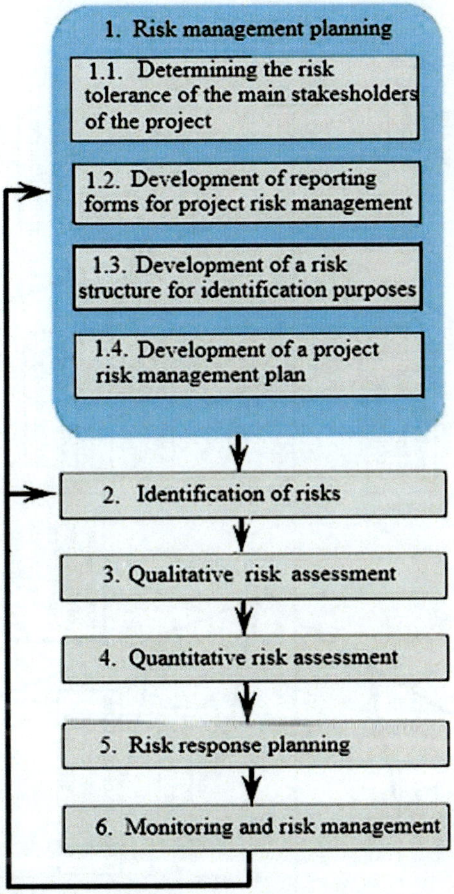

Fig. 5.3 Generalized risk management planning algorithm for Arctic oil and gas projects

- Methods of collecting information, including expert methods of brainstorming and Delphi, surveys among employees participating in the project and among relevant experts, SWOT analysis, etc.
- Analysis of risk checklists developed on the basis of already available information on already implemented projects
- Analysis of scenarios, hypotheses, and assumptions of the project allows you to identify risks arising from inaccuracies, incompatibility, or incompleteness of project assumptions
- Risk mapping using diagrams, including Ishikawa diagrams with an assessment of the causes, dependencies, and sequences of project risks (Fig. 5.4)

The result of the identification process is a risk register, which, in addition to describing the risks themselves, also contains the main causes of their occurrence (Table 5.1).

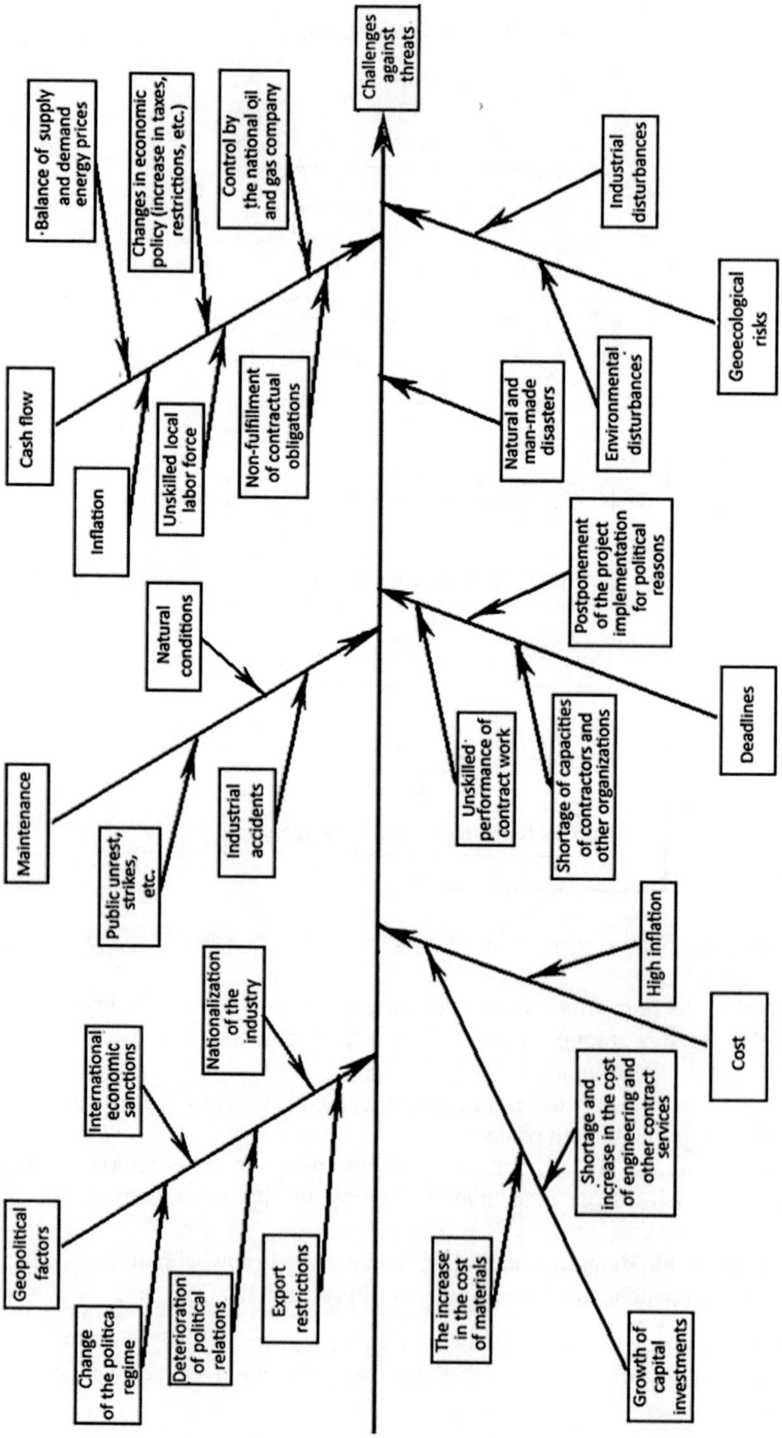

Fig. 5.4 Ishikawa Diagram for risk identification and management of Arctic oil and gas projects

Table 5.1 Register of the main risks of development of oil and gas projects in the Arctic region

Risk characteristics		Risk classification			Probability of occurrence	Effects	
Name	Factors	Country	Type	Scale		Type	Degree of influence
Economic risks							
Falling oil/gas prices below the expected level	1. Changing the state of the world economy 2. High volatility of energy prices 3. Fierce competition in the sale of energy resources in certain markets 4. The emergence of new energy sources		Economic	Portfolio	Mean	Deadlines Cost Content /quality Cash flow	– – Mean High
A significant increase in the cost of entering the project as the stages of its implementation are completed	1. Unwillingness of existing project participants to withdraw from it at late stages of implementation 2. Reducing risks and uncertainty about the success of the project as it is implemented		Economic	Portfolio	High	Deadlines Cost Content /quality Cash flow	– High – –
The increase in the risk of capital investments for the implementation of the project	1. The increase in the cost of contractors' services, materials 2. Unskilled (low-quality, untimely, overspending) performance of contract work 3. Errors in capital expenditure planning and budgeting		Economic	Portfolio	High	Deadlines Cost Content /quality Cash flow	– High Mean Low

(continued)

Table 5.1 (continued)

Risk characteristics		Risk classification			Probability of occurrence	Effects	
Name	Factors	Country	Type	Scale		Type	Degree of influence
Risk of shortage and increase in the cost of engineering services, contractors, materials	1. The imbalance of supply and demand of these services 2. Inflationary price growth 3. Global growth of prices for basic resources (metals, building materials, etc.)		Economic	Portfolio	Mean	Deadlines Cost Content /quality Cash flow	Mean High – Mean
The risk of the project implementation period due to the absence, insufficiency, or increase in the cost of financial resources	1. Financial position of the participating companies of the project 2. The crisis situation in the international financial markets 3. High growth in the implementation of projects due to uncertainty about the pace of development of the world economy and demand for energy resources		Economic	Portfolio	Mean	Deadlines Cost Content /quality Cash flow	High High High High
Geopolitical risks							
Risk of nationalization of assets	1. Political instability. 2. Changing the economic policy of the state 3. Change of the political regime	Russia, USA, China, Finland, Norway, Sweden, Canada, Japan, Korea, Malaysia	Geopolitical	Sub-portfolio	Low	Deadlines Cost Content /quality Cash flow	– – Low Mean

Risk characteristics		Risk classification			Probability of occurrence	Effects	
Name	Factors	Country	Type	Scale		Type	Degree of influence
The risk of deterioration of the activities of foreign companies in the country	1. Changing the economic policy of the state 2. Deterioration of the political relations of the countries 3. The state's desire to maximize revenue from the industry	Russia, USA, China, Finland, Norway, Sweden, Canada, Japan, Korea, Malaysia	Geopolitical	Sub-portfolio	Low	Deadlines Cost Content /quality Cash flow	Mean Low – Mean
Restriction of export supplies from the country	1. Domestic energy demand and introduction of export restrictions 2. The introduction of international sanctions against the country implementing the project, including a ban on the import of energy resources from this country	Russia, USA, China, Finland, Norway, Sweden, Canada, Japan, Korea, Malaysia	Geopolitical	Sub-portfolio	Mean	Deadlines Cost Content /quality Cash flow	– – – Mean
The risk of the outbreak of hostilities in the project implementation region, terrorist attacks	1. Instability of the political regime 2. Threat of military invasion	Russia, USA, Finland, Norway, Sweden, Canada	Geopolitical	Sub-portfolio	Low	Deadlines Cost Content /quality Cash flow	Low Low Low Low

(continued)

Table 5.1 (continued)

Risk characteristics		Risk classification			Probability of occurrence	Effects	
Name	Factors	Country	Type	Scale		Type	Degree of influence
Problems related to the transportation of energy carriers by sea	1. The introduction of international sanctions against the country implementing the project, including a ban on the import of energy resources from this country	Russia, USA, China, Finland, Norway, Sweden, Canada, Japan, Korea, Malaysia	Geopolitical	Sub-portfolio	High	Deadlines Cost Content /quality Cash flow	High High Mean High
Problems related to the transportation of energy carriers via onshore pipelines	1. The introduction of international sanctions against the country implementing the project, including a ban on the import of energy resources from this country	Russia, USA, China, Finland, Norway, Sweden, Canada, Japan, Korea, Malaysia	Geopolitical	Sub-portfolio	Mean	Deadlines Cost Content /quality Cash flow	Mean Mean Mean High
Aggravation of international competition for oil and gas assets in various regions of the world	1. Strong positions of the world's largest oil and gas companies, American and European, in a number of resource-rich countries 2. Active participation of Chinese capital in competition in the oil and gas assets market		Geopolitical	Sub-portfolio	High	Deadlines Cost Content /quality Cash flow	– High – –

Risk characteristics		Risk classification			Probability of occurrence	Effects	
Name	Factors	Country	Type	Scale		Type	Degree of influence
Insufficient state support for the activities of Russian companies and problems in obtaining control over the activities of enterprises in other countries	1. Legislative restrictions on the rights of foreign enterprises 2. Subjective negative attitude toward Russian companies		Geopolitical	Sub-portfolio	High	Deadlines Cost Content /quality Cash flow	Mean High – High
Natural risks							
Risks associated with adverse environmental conditions	1. The ice situation that exists during most of the year 2. Strong winds throughout the year 3. The presence of floating ice and stamukh 4. Abrasion and erosion of the coastline		Natural	Portfolio	High	Deadlines Cost Content /quality Cash flow	High High High High
Negative impact on the work of the staff	1. Extreme weather conditions in the project implementation region 2. Conditions of permafrost, polar night, etc.		Natural	Portfolio	High	Deadlines Cost Content /quality Cash flow	High High High High

(continued)

Table 5.1 (continued)

Risk characteristics		Risk classification			Probability of occurrence	Effects	
Name	Factors	Country	Type	Scale		Type	Degree of influence
Geoecological risks							
Negative reversible effect of anthropogenic altered polar environment on the operation of oil and gas complex enterprises	1. Thermokarst processes subsidence and sinkholes, deformation of foundations and foundations, provoked accelerated erosion and solifluction, threatening both linear and areal objects		Natural	Portfolio	High	Deadlines Cost Content /quality Cash flow	Mean High Mean Mean
Technological risks							
The risk of an accident (explosion) at the plant (transport) with damage to the tangible and intangible assets of the enterprise, as well as the external environment	1. Technical malfunctions of the equipment 2. Incorrect design of the construction of mechanisms 3. Unskilled personnel 4. Non-compliance with safety regulations 5. Violation of the rules of operation of the equipment 6. Spontaneous combustion of raw materials and products of its processing 7. Violation of fire safety regulations		Technological	Portfolio	Low	Deadlines Cost Content /quality Cash flow	High High High High

Risk characteristics		Risk classification			Probability of occurrence	Effects	
Name	Factors	Country	Type	Scale		Type	Degree of influence
Disruption of construction and installation work deadlines	1. Lack of accurate estimates 2. Lack of previous experience in polar conditions 3. Lack of construction precedents 4. Mistakes in project design 5. Lack of qualified personnel Unskilled (low-quality, untimely, overspending) performance of contract work		Technological	Portfolio	High	Deadlines Cost Content /quality Cash flow	High Low Low Mean
Social risks							
Possible negative opinion of the public about the implementation of the project	1. Negative impact on the environment 2. Ignoring legal norms and rules 3. Non-compliance with local customs 4. The use of social engineering by competitors		Social	Sub-portfolio	Mean	Deadlines Cost Content /quality Cash flow	High Low Low Mean
Unskilled personnel (among the local population)	1. Insufficient number of locally trained personnel 2. Requirements for the involvement of local staff by regional authorities		Social	Sub-portfolio	Mean	Deadlines Cost Content /quality Cash flow	Mean Mean Mean Mean

(continued)

Table 5.1 (continued)

Risk characteristics		Risk classification			Probability of occurrence	Effects	
Name	Factors	Country	Type	Scale		Type	Degree of influence
Public unrest, strikes, and other expressions of discontent in the project implementation region	1. Political and social instability 2. Poverty of the population and high unemployment 3. Dissatisfaction of the population with the distribution of income from the extraction and export of resources		Social	Sub-portfolio	Low	Deadlines Cost Content /quality Cash flow	Low Low – Mean

At the same time, the following are the most significant risks of Arctic oil and gas projects:

- Falling energy prices below a certain level
- A significant increase in the cost of entering into projects as it goes through the stages of its implementation
- Risk of capital investment growth for project implementation
- Risk of shortage and increase in the cost of engineering services, contractors, and materials
- Risk of nationalization of assets
- Risk of deterioration of the activities of foreign companies in the country
- Risks associated with the transportation of energy carriers by sea
- Risks associated with the transportation of energy carriers by land
- Risks associated with the lack of necessary technologies for the implementation of the project, which is especially important in the conditions of the Arctic environment
- Geoecological risks caused by the interdependent impact of an industrial project on the environment and the anthropogenic changed environment on the industrial facilities themselves

It should be noted that risk identification should be carried out throughout the life cycle of an oil and gas Arctic project for a number of reasons:

1. During the implementation of the project, new risks that were not identified at the planning stage may arise.
2. The probability and impact of risks on the objectives of these projects may change during their implementation.

The main problems of the identification process are the incompleteness and inaccuracy of the risks identified during the implementation of the project. Failure to detect the risks of projects that affect the course of the project can lead to a significant loss of the effect of investments.

A generalized algorithm for identifying risks of Arctic oil and gas projects is shown in Fig. 5.5.

After identifying the risks of Arctic oil and gas projects, assessments of the most significant of them are carried out at the qualitative and quantitative levels.

5.1.3 Qualitative Risk Assessment of Arctic Oil and Gas Projects

The need for a qualitative assessment of Arctic oil and gas projects is explained by the following main reasons:

- The need to simultaneously take into account a huge amount of initial information of an economic, geopolitical, geoecological, institutional, economic, man-made, and other nature, for a reasonable choice of the development option of the Arctic business of an oil and gas company in conditions of uncertainty
- Physical impossibility of taking into account all risk factors during quantitative calculations

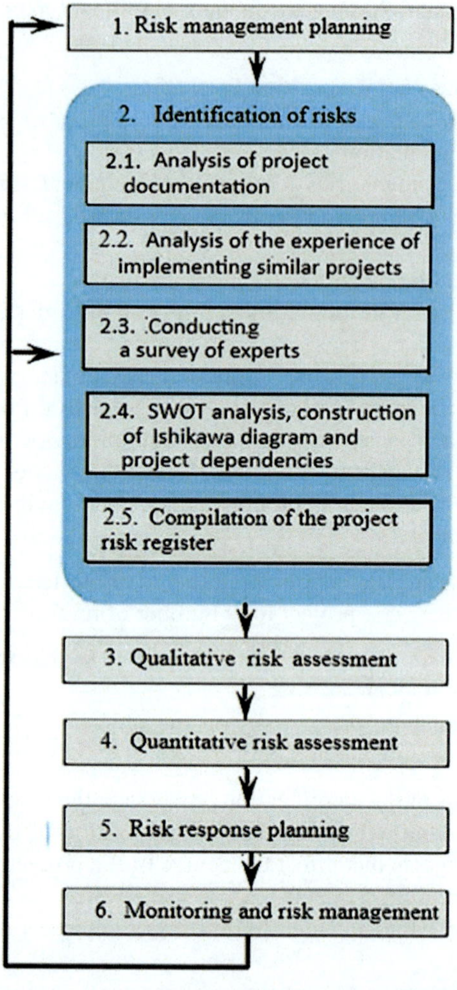

Fig. 5.5 Generalized risk identification algorithm for Arctic oil and gas projects

The purpose of a qualitative risk assessment is to prioritize identified risks. The selection of the most significant risk factors subsequently simplifies the procedures for their quantitative analysis and planning of response to them. Priorities for identified risks are determined based on the probability of occurrence of risk events and the degree of risk impact on the success criteria (goals) of projects. At the same time, the limitations of Arctic oil and gas projects (time and cost frames, attitude to risk) are taken into account. The main indicators of the success of such projects for the purposes of risk analysis are the timing, cost, quality, and cash flow of projects.

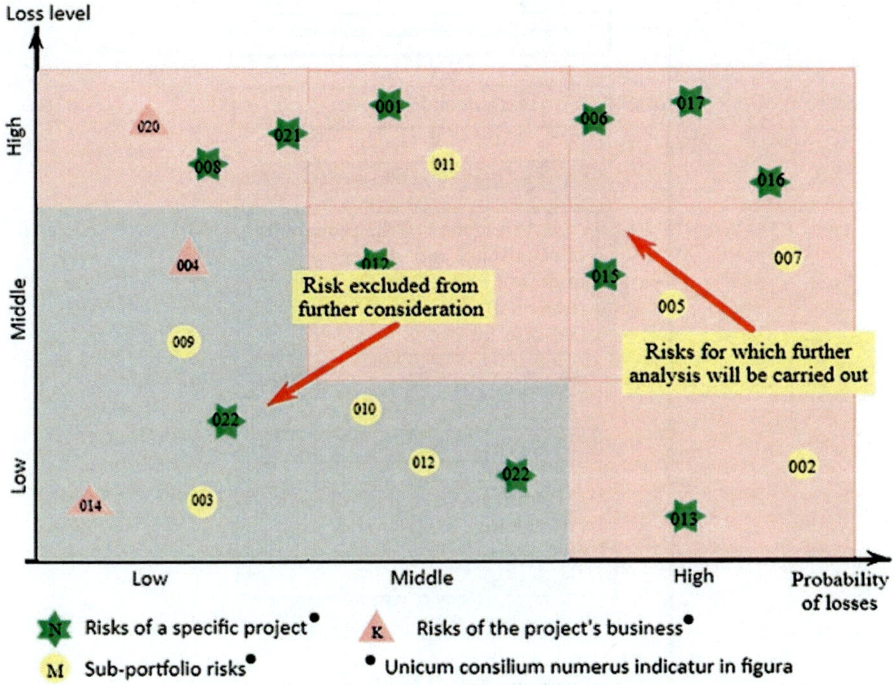

Fig. 5.6 Risk map of Arctic oil and gas projects in the coordinates “probability–impact”

The probability of occurrence of risk events and impact levels are assessed separately for each identified risk by expert methods, since information about the manifestation of risks for past projects is insufficient in the absolute majority of such projects. The assumptions used to determine risk levels are documented.

At the next step, a risk map is built in the coordinates “probability–impact” (Fig. 5.6), after which its importance is calculated for each identified risk as the product of the probability of its occurrence and impact on the project goals. The risks are then ranked by importance.

For risks of high importance, a quantitative assessment is subsequently carried out and response strategies are developed. For risks that do not have a significant impact on the project, as a rule, a passive observational strategy (monitoring) is used.

The main results of a qualitative risk assessment are as follows:

- Ranked list of risks
- Risk map
- Lists of risks requiring response and monitoring
- List of significant risks that require additional quantitative analysis

A generalized algorithm for qualitative risk assessment is presented in Fig. 5.7.

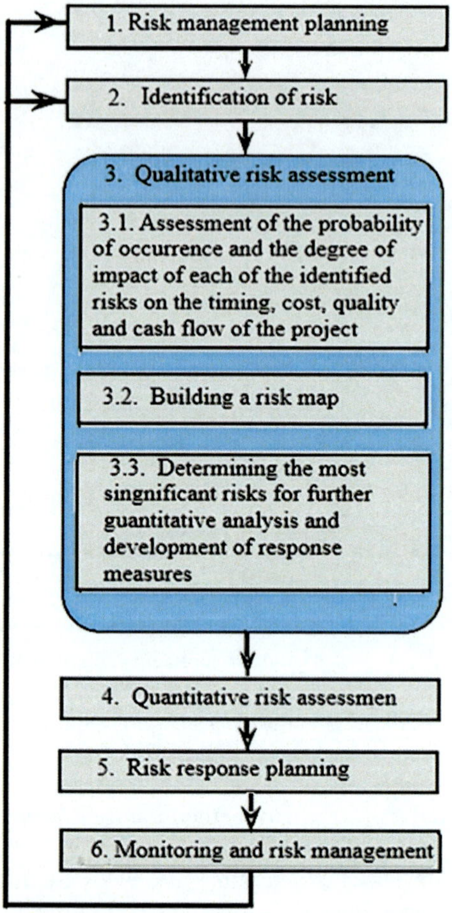

Fig. 5.7 Generalized algorithm for qualitative risk assessment

5.1.4 Quantitative Risk Assessment of Arctic Oil and Gas Projects

Quantitative assessment of project risks is carried out only in relation to those risks that were recognized as significant at the previous stage.

The objectives of quantitative risk analysis of Arctic oil and gas projects may be as follows:

- Assessment of the probability of achieving the project objectives in terms of cost, timing, or content, taking into account the impact of risks
- Identification of risks that require the most attention from managers by quantifying their relative contribution to the overall risk of the project
- Development of a model for the subsequent determination of the best management decisions to respond to risks

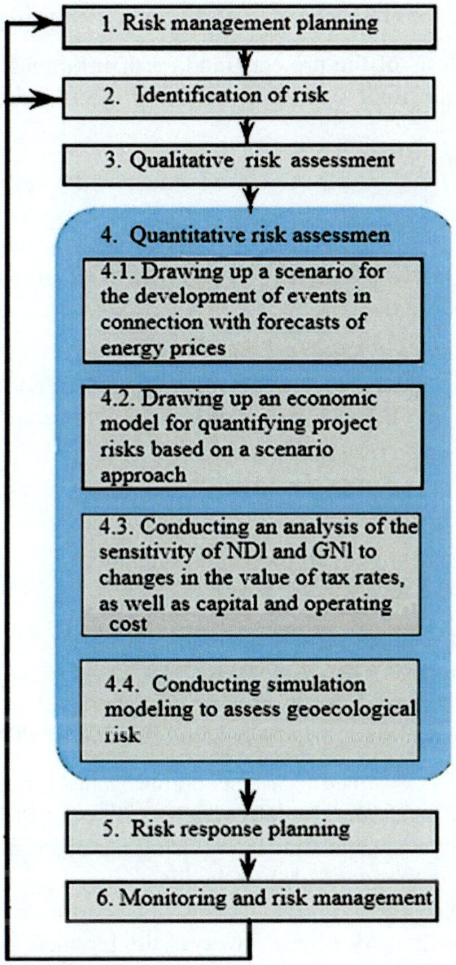

Fig. 5.8 Generalized algorithm for quantitative risk assessment of the projects under consideration

Sensitivity analysis is one of the simplest and most common tools for quantifying the risks of Arctic oil and gas projects. It is advisable to analyze the sensitivity of the NDI and GNI indicators to changes in the values of tax rates, as well as capital and operating costs.

For Arctic projects, it is extremely important to carry out a quantitative assessment of geoecological risks, examples of which are given in Chap. 3.

A generalized algorithm for quantifying the risks of the projects under consideration is presented in Fig. 5.8.

The result of the quantitative risk assessment of Arctic oil and gas projects is an updated risk register, an economic and environmental model of the project, which can be used to assess the effectiveness of risk response measures.

The following sections are subject to updating in the risk register:

- Probabilistic analysis of the project (random distributions of cost, timing, and cash flow from the implementation of the projects under consideration are presented)
- The probability of achieving the project goals

5.1.5 Risk Response Planning for Arctic Oil and Gas Projects

Risk response planning for Arctic oil and gas projects is a process of developing strategies and measures aimed at strengthening positive trends and reducing threats to the project objectives. Planned risk response activities must meet the following requirements:

- Efficiency of use
- Timeliness of application
- Realistic and consistent with all project participants
- Compliance with the severity of the risk

The main problem solved by management during the implementation of this process is the formation of a list of measures to respond to individual risks and the selection of the best combinations of these measures.

The following types of basic risk response strategies are distinguished:

- Risk avoidance. It is assumed to use such approaches that exclude the negative impact of risk, for example, the refusal of the services of this counterparty. In other cases, risk avoidance is aimed at revising the objectives of the project, for example, expanding the scope of the schedule.
- Transfer of risk. This involves transferring all or part of the risk to another person. As a rule, the transfer of risk involves the payment of a premium to this person. During the transfer, the risk is not eliminated, but shifted to the other side. Risk transfer instruments for Arctic oil and gas projects include, in particular, insurance, hedging, use of guarantees, real options, etc.
- Risk reduction. It is assumed to reduce the probability of a threat and/or reduce the consequences of a negative event to an acceptable size. First, measures are being developed to reduce the likelihood of risky events, for example, conducting more tests. If it is not possible to reduce the risk to an acceptable level, measures should be aimed at reducing its consequences.
- Passive or active risk taking. The passive risk-taking strategy assumes management's actions at its own discretion in the event of risk events. It is used to respond to insignificant risks.

When planning Arctic oil and gas projects, a contingency response plan is also being developed in accordance with the strategy for responding to them. These measures should be carried out in case of occurrence of risky events.

The results of the risk response process are as follows:

- Updated risk register, which includes information on selected risk response strategies and activities
- Updated project management plan, including its budget and schedule
- Contractual agreements concerning risk response measures (insurance contracts, liability of the parties in purchase and sale contracts, provision of services, etc.)
- Extensions to the risk register, which relate to selected and agreed risk response strategies
- Budget and planned operations required to implement risk response measures
- The budget and reserves for unforeseen circumstances, plans in case of their occurrence, as well as the conditions and events under which they are put into effect
- Backup plans if the selected risk response measures have proved ineffective
- Residual risks that exist after the implementation of risk response processes
- Secondary risks that have arisen as a result of risk response measures

The main problem of this process is the selection and justification of effective risk response measures.

A generalized algorithm for planning response to the risks of Arctic oil and gas projects is presented in Fig. 5.9.

5.1.6 Monitoring and Risk Management of Arctic Oil and Gas Projects

Monitoring and risk management of Arctic oil and gas projects consists of the following processes:

- Identification, assessment, and development of plans to respond to risks that arise again during the implementation of Arctic oil and gas projects
- Monitoring the status of identified risks
- Execution of risk response operations and assessment of their effectiveness in case of realization of relevant risks

The following main methods and tools are used in the process of monitoring and risk management:

- Risk review – identification of new and revision of known risks, their qualitative and quantitative analysis, as well as planning of measures to respond to newly emerging risks; this process is carried out on a regular basis
- Risk audit – assessment of the effectiveness of risk management processes of the projects under consideration
- Analysis of trends and deviations – forecasting potential deviations in the course of the project by timing and cost

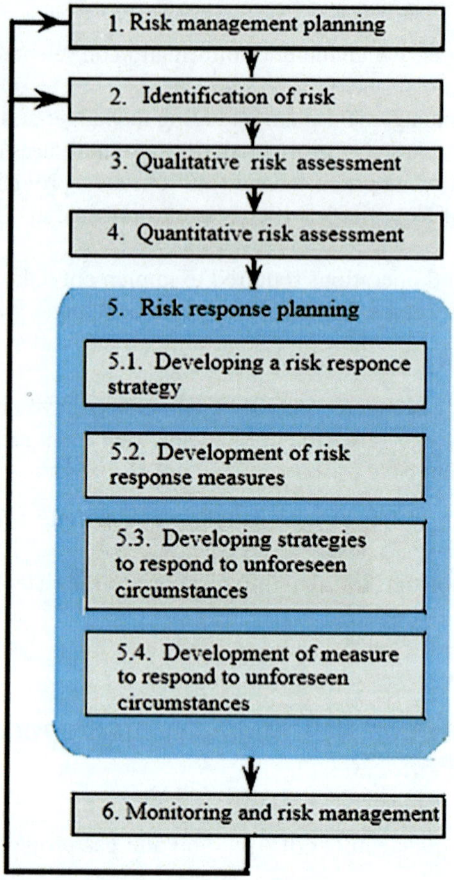

Fig. 5.9 Generalized algorithm for planning response to risks of Arctic oil and gas projects

- Comparison of the technical results of the project with the planned ones – contributes to making forecasts about the degree of success in achieving the project goals
- Analysis of residual reserves – comparison of the volume of residual reserves of each type with the volume of residual risks
- Meetings on the current state of the project

The result of the risk monitoring and management process is an updated risk register. In case of new risks, updates may affect all registry keys. In some cases, only the project management plans need to be updated.

The main problem of this process is the correct assessment of residual reserves and the choice of effective measures to respond to newly emerging risks.

A generalized algorithm for monitoring and risk management of Arctic oil and gas projects is presented in Fig. 5.10.

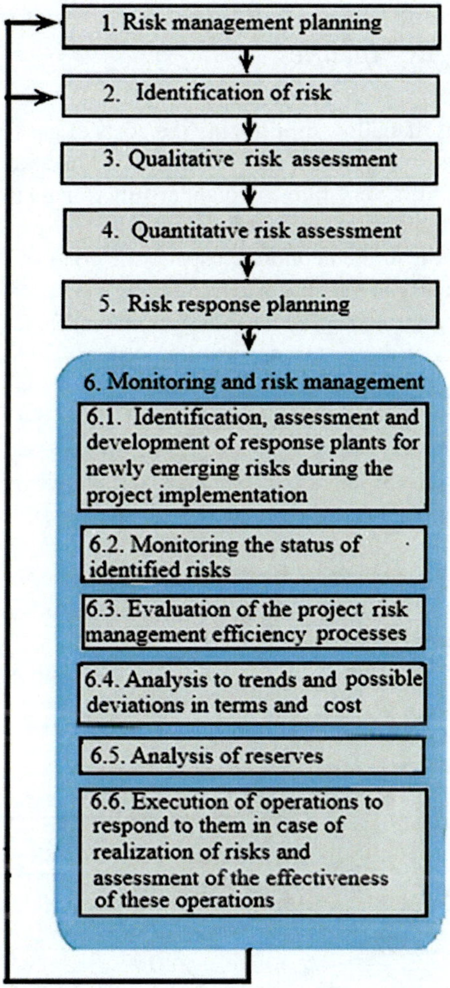

Fig. 5.10 Generalized algorithm for monitoring and risk management of Arctic oil and gas projects

5.2 Management of Geoecological Risks on Complex Main Gas Pipelines

Chapter 3 describes in detail the models for calculating geoecological risk parameters for both marine and coastal Arctic ecosystems in the impact zones of the gas industry. The different probability of occurrence of risks in different scenarios of development, both planned and already implemented projects, was shown. At the same time, as shown earlier (Bashkin & Kazak, 2015), the probability of geoecological risks is significantly higher in the impact zones of main gas pipelines. Accordingly, it is necessary to consider risk management techniques in these conditions.

5.2.1 *Management of Geoecological Risks on Linear Parts of Main Gas Pipelines*

Mathematical models that allow minimizing the costs of gas transportation through complex gas transmission networks have been studied in detail. However, in practical implementations, it is very rare to consider functionals that take into account, along with technical and economic indicators, geoecological and environmental components. Of course, in the economic aspect, the costs of reducing geoecological risks are significantly lower than the total energy costs. Nevertheless, the optimization of gas flows through gas transmission networks, taking into account the economic assessment of the gas transportation system (GTS) impact on the environment, will make it possible to develop solutions for managing geoecological risks. This makes it possible to reduce the costs of direct gas losses resulting from its leaks during transportation, storage, and repair work. In addition to the obvious economic damage from these losses, this approach allows us to take into account the reduction of indirect losses associated with the possible sale of greenhouse gas quotas.

Let us consider the described methodological approaches and mathematical models that allow us to implement the principles outlined above (Samsonov et al., 2007a, b; Rusakova et al., 2009; Kazak & Bashkin, 2015a, b). Figure 5.11 shows an aggregated schematic diagram of the gas transmission system.

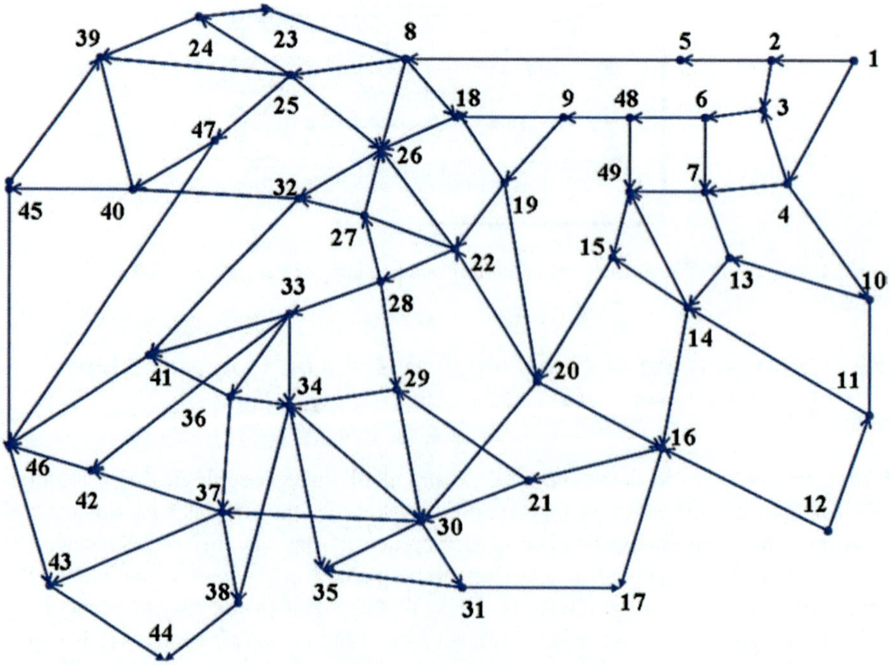

Fig. 5.11 Schematic aggregated diagram of the gas transmission system

The degree of aggregation of the system was chosen based on existing zones with different prices for the sale of natural gas, so that for GTS sections, this indicator at the inlet and outlet was different.

With current approaches, planned gas flows are calculated by maximizing the functionality:

$$F = \sum_{i=1}^{n} Q_i \cdot v_i \rightarrow \min \tag{5.1}$$

where

Q_i is the flow along the *i-th* arc
v_i is the weight of the *i-th* arc

Optimization of the functionality is carried out under restrictions:

$$Q_i \leq M_i, \tag{5.2}$$

where M_i is the maximum performance of the gas pipeline section.

At the same time, the equality from Kirchhoff's law is fulfilled for each node:

$$\sum_{i \in j} \pm Q_i \pm S_j = 0 \tag{5.3}$$

where S_j is the gas supply (+) or consumption (−) in the *j-th* node of the system.

The weight of the *i-th* arc for this problem is determined from the ratio

$$v_i = T_i, \tag{5.4}$$

where T_i is the unit costs of gas transportation.

In fact, we are talking about calculating annual gas flows across all sections of the gas transmission system, provided that transportation costs are minimized and gas supply obligations are fulfilled to consumers. The value T_i, in turn, is a function of the gas flow Q_i; therefore, the optimization of the functional (1) is carried out iteratively. Methods of constructing and optimizing gas flows through complex gas transmission systems and gas distribution networks have been well studied for various classes of problems (Chernyaev et al., 1991; Kazak et al., 1985, 1989). In this regard, we will not dwell on them in detail, and we will proceed to the formation of a functional for optimizing gas flows taking into account the environmental component.

For each site, the economic component of the assessment of geoecological consequences includes the following:

- Determination of the volume of probable gas emissions in accidents and leakage based on statistical models
- Determination of gas bleed volumes during scheduled repair works based on optimization of their schedules
- Determination of the volume of gas consumed for fuel needs based on minimizing energy consumption at compressor stations

Thus, the weights of the arcs in the minimized functional are determined from the ratio:

$$v_i = T_i + E_i, \tag{5.5}$$

where E_i is the economic component of the environmental impact on the environment during the year on the *i-th* section of the gas transmission system.

Let us explain the above on the simplest numerical example (Fig. 5.12).

Suppose that node S produces 400 billion m³/year. Accordingly, 400 billion m³/year is consumed in node W (gas costs for own needs are tied to node W). Additional data required for the calculation behavior are presented in Table 5.2.

Let option 1 be understood as the calculation of the gas flow without taking into account the environmental component, and option 2, respectively, taking into account the environmental component. Then for option 1 we have the following:

Weight of the 1st arc v_i = 2*200/100 = 4.00 \$/thousand m³
Weight of the 2nd arc v_i = 2*210/100 = 4.20 \$/thousand м³

By optimizing the restrictions on the maximum performance of the arc (section of the gas pipeline), we obtain gas flows (Table 5.2).

For variant 2 we have:

Weight of the 1st arc v_i = 2*200/100 + 0.25*200/100 = 4.50 \$/ thousand м³
Weight of the 2d arc v_i = 2*210/100 + 0.10*210/100 = 4.41 \$/ thousand м³

The results of optimization, taking into account, the restrictions on the maximum performance of the arc (section of the gas pipeline) due to the environmental component are presented in Table 5.3.

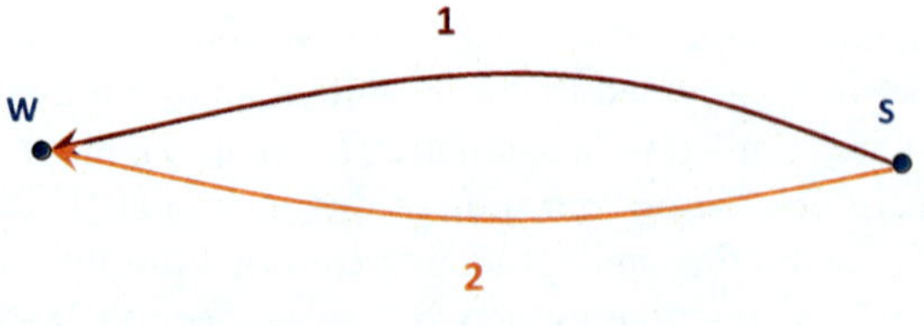

Fig. 5.12 Example of a GTS scheme for calculating gas flows

Table 5.2 Necessary data for the calculation

Arc number	Arc length, km	Max. capacity, billion m³/year	Unit costs (T) \$/(100 km* thousand m³)	Ecological component (E) \$/(100 km* thousand m³	Flow option 1, billion m³/year	Flow option 2, billion m³/year
1	200	210	2	0.25	210	170
2	210	230	2	0.10	190	230

Table 5.3 Results of optimal redistribution of gas flows

Plot number	Node numbers	Gas flow, million cubic meters excluding the environmental component	Gas flow, million cubic meters taking into account the environmental component	Risk level
1	2	3	4	5
1	1–2	510600	519016	L
2	2–3	423000	421842	M
3	4–3	2700	1844	H
4	1–4	49700	41284	H
5	2–5	72000	81574	L
6	3–6	397700	395686	M
7	4–7	13080	11062	H
8	6–7	92300	84972	H
9	4–10	21620	16079	H
10	10–11	11020	6328	H
11	10–13	2000	1150	H
12	11–14	5170	4100	H
13	12–11	5750	9371	L
14	7–13	18650	16137	H
15	5–8	184500	194074	L
16	6–48	281600	286914	M
17	7–49	73730	66896	H
18	13–14	9750	6388	H
19	14–15	5500	3836	H
20	12–16	61150	57529	H
21	9–18	23700	23740	M
22	9–19	220300	216806	M
23	15–20	70130	69596	M
24	20–16	5040	3183	H
25	16–17	26100	29848	L
26	8–23	21400	23241	L
27	8–25	119920	115274	H
28	8–26	25080	37458	L
29	8–18	0	0	b
30	18–26	6570	5578	H
31	19–18	8370	7538	H
32	19–22	200990	200541	M

(continued)

Table 5.3 (continued)

Plot number	Node numbers	Gas flow, million cubic meters excluding the environmental component	Gas flow, million cubic meters taking into account the environmental component	Risk level
33	19–20	4640	2626	H
34	20–22	13090	9671	H
35	16–21	73710	62520	H
36	20–30	43040	45768	L
37	21–29	21630	17402	H
38	21–30	38280	31318	H
39	31–17	25040	21292	H
40	22–26	22300	16330	H
41	22–27	23200	26054	L
42	22–28	166380	165628	M
43	24–23	4500	2659	H
44	25–24	17000	14020	H
45	25–26	3220	2516	H
46	27–26	9130	4418	H
47	28–27	10630	3063	H
48	28–29	23090	20031	H
49	29–30	16120	17875	L
50	30–31	39640	32281	H
51	26–32	0	0	[b]
52	27–32	13100	13099	M
53	28–33	120960	130834	L
54	29–34	25200	16158	H
55	30–35	22500	27558	L
56	30–37	20300	20174	M
57	33–34	3000	6275	L
58	33–36	52300	52299	M
59	34–35	4000	2553	H
60	36–34	6870	5004	H
61	34–38	33410	27172	H
62	36–37	36030	42268	L
63	37–38	5320	11558	L
64	24–39	8000	6861	H

(continued)

Table 5.3 (continued)

Plot number	Node numbers	Gas flow, million cubic meters excluding the environmental component	Gas flow, million cubic meters taking into account the environmental component	Risk level
65	25–47	94400	93439	**M**
66	32–41	7400	7031	**H**
67	33–41	14200	20802	**L**
68	36–41	16100	11727	**H**
69	33–42	45060	45059	**M**
70	37–42	20300	20174	**M**
71	37–43	28910	28910	**M**
72	38–44	13630	13630	**M**
73	43–44	32070	32070	**M**
74	40–39	10310	11462	**L**
75	40–45	33710	31965	**H**
76	41–46	14600	16459	**L**
77	42–46	63360	63233	**M**
78	46–43	3560	3560	**M**
79	45–39	1990	1978	**M**
80	45–46	19420	17688	**H**
81	14–16	30120	28156	**H**
82	47–40	57120	56159	**M**
83	47–46	26080	26080	**M**
84	48–9	252600	249146	**M**
85	48–49	11500	20268	**L**
86	14–49	2500	1696	**H**
87	49–15	88230	89360	**M**
88	30–34	14000	13948	**M**
89	31–35	3900	289	**H**
90	32–40	0	367	**L**
Functional F[a]		**2110**	**2090**	**M**

[a]The value of the functional F is determined from the ratios (1–4); the gas flows correspond to the values in the 4th and 5th columns of the table; the weights of the GTS sections are calculated by the formula (5)

[b]The plot is under repair

Thus, from the calculation results obtained, it can be seen that in option 1 the first section of the system was fully loaded, and in option 2, respectively, the second. The simplest example presented shows that the environmental component can significantly affect the redistribution of gas flows.

In relation to the system under consideration (Fig. 5.12), practical calculations have been carried out, the results of which are presented in Table 5.3.

The value of the environmental risk level in Table 5.3 is defined in the following range: "low" (**L**), "mean" (**M**), and "high" (**H**). Risk levels are assessed based on changes in the gas flow for each site. For the sites where a significant increase in the gas flow was obtained because of optimization taking into account the environmental component, in relation to a similar value calculated without taking it into account, it is assumed that the level of geoecological risk at this site is "low". In this case, in relation to other sites, the consequences of the operation of the gas pipeline on the environment will be lower. These consequences are estimated in value terms and relate to the unit volume of gas pumping at the GTS site. Similarly, in the case of a significant decrease in gas flow (column 4, Table 5.3) in relation to its value without taking into account the environmental component (column 3, Table 5.3), the level of geoecological risk is assessed as "high". For the remaining sites, the risk level is defined as "mean." In the above calculations, the range of the "mean" level of geoecological risks at the site is assumed as 1.5% of the gas flow value calculated without taking into account the geoecological component.

The results obtained allow us to identify areas where geoecological risks are relatively small. The risk level for such sites is highlighted in green in Table 5.3. At the same time, the calculations carried out make it possible to identify areas where there is an increased (relative to the average value for the entire system) level of the impact of emissions from main gas pipelines on the ecological state of the environment (in Table 5.3, these areas are highlighted in red). Thus, for each disadvantaged section of the GTS from the above list, it is advisable to provide measures to reduce environmental risks on it. In addition, the proposed approach makes it possible to assess the total geoecological risks, both for the entire GTS system and individual subsystems. Table 5.3 presents a weighted average assessment of the geoecological risk for the entire system, which shows that the environmental impacts of the emissions of this GTS system are still at an average acceptable level (functional F).

5.2.2 *Geoecological Risk Management By Optimizing the Placement and Operation of Compressor Stations on Main Gas Pipelines*

Management of geoecological risks in the operation of compressor stations, as elements of the gas transmission system, involves the development of programs to control and reduce emissions of pollutants. The main directions that allow the reduction of emissions of NO_x, CO, CH_4, and CO_2 include the following:

- Development and implementation of new types of low-toxic GPU
- Modernization of combustion chambers operated by GPU

- Research on the influence of regime and operational factors
- Studies of the processes of dispersion and transformation of emissions in the atmosphere
- Monitoring of emissions and the state of the atmosphere
- Creation of catalytic combustion chambers (NO_x emission less than 10 mg/m^3)

The introduction of new low-emission technologies in gas transportation is one of the most important methods of environmental risk management, especially in places where the environmental impact has reached a critical level.

Reduction of emissions of the most harmful gases during gas transportation (nitrogen, carbon, and sulfur oxides – the latter is subject to the presence of hydrogen sulfide in natural gas, which is typical for the southern Astrakhan and Orenburg fields, Russia) and reduction of environmental impact with a corresponding reduction in geoecological risks occur within the framework of reducing energy costs.

Let us consider mathematical models and algorithms that allow minimizing energy consumption on a section of a gas pipeline with several compressor stations, provided it is fully loaded.

The main method of calculating the operation mode of the supercharger is the use of its compression ratio characteristics – $\varepsilon(Q,\tilde{n})$, polytropic efficiency - $\eta_{пол}(Q,\tilde{n})$ and the internal effective power of the supercharger $N_i(Q,\tilde{n})$, which establishes an empirical relationship between the specified parameters of the supercharger, its volumetric capacity, and the shaft speed.

In accordance with the existing technical rules, the use of third-degree polynomials approximating the characteristics from the volumetric reduced to the nominal performance is regulated for these purposes.

Compression ratio:

$$\varepsilon = a_\varepsilon + b_\varepsilon \cdot q_{\text{пр}} + c_\varepsilon \cdot q_{\text{пр}}^2 + d_\varepsilon \cdot q_{\text{пр}}^3 \tag{5.6}$$

Polytropic efficiency:

$$\eta_{пол} = a_\eta + b_\eta \cdot q_{\text{пр}} + c_\eta \cdot q_{\text{пр}}^2 + d_\eta \cdot q_{\text{пр}}^3 \tag{5.7}$$

Internal effective power of the supercharger

$$N_{\text{пр}} = a_N + b_N \cdot q_{\text{пр}} + c_N \cdot q_{\text{пр}}^2 + d_N \cdot q_{\text{пр}}^3 \tag{5.8}$$

where a, b, c, d are the coefficients of approximation of the corresponding reduced characteristic and $q_{\text{пр}}$ is the volumetric reduced flow:

$$q_{\text{пр}} = Q_{\text{об}} \frac{1}{\tilde{n}} \tag{5.9}$$

where, $Q_{об}$ is the volumetric reduced flow rate at the inlet of the supercharger, m³/min or m³/s and $\tilde{n}$ is the relative revolutions; $(\tilde{n} = \frac{n}{n_н})$

$$q_{тг} = q^0_{тг}\left(0.75\cdot\frac{N}{N_н} + 0.25\cdot\sqrt{\frac{T_{воз}}{T_{нв}}\cdot\frac{p_{воз}}{p_{нв}}}\right)\cdot K_{тг}\cdot K_{\tilde{n}} \tag{5.10}$$

where, $q^0_{тг} = \frac{3,6\cdot N_н}{\eta_{ном}\cdot H^*_{сн}}$ is the nominal fuel gas consumption, thousand m³/hour; $N_н$ is the rated power, kW; $\eta_{ном}$ is the nominal efficiency of a gas turbine plant; $H^*_{сн}$ is the nominal lowest specific volumetric heat of combustion of fuel (assumed to be equal to 34,500 KJ/m³); $T_{воз}$ is the estimated atmospheric air temperature [^0K]; $T_{нв}$ is the nominal air temperature [^0K]; $p_{воз}$ is the design air pressure [MPa]; $p_{нв}$ is the nominal air pressure [MPa]; $K_{тг}$ is the coefficient of technical condition; and $K_{\tilde{n}}$ is usually taken into account in the composition $K_{тг}$.

Fuel gas consumption for the billing period is determined by the following formula (Sardanashvili, 2005):

$$Q^г_{тг} = q_{тг}\cdot\tau\cdot 10^{-3} \tag{5.11}$$

Where, τ is the time of the billing period.

The initial data for this subsystem will be the characteristics necessary for the calculation of active elements – compressor stations and passive – linear sections (Samsonov et al., 2007b; Rusakova et al., 2009). For the unambiguity of the problem statement, it is also necessary to set boundary conditions. For a chain-type corridor, such conditions will be the pressure and gas flow at the inlet. The task is to form optimal control (operating modes of all compressor stations of the subsystem) in order to minimize the cost of pumping gas, i.e., minimization of the total cost of gas consumed by gas turbine installations and electricity on electric drive installations.

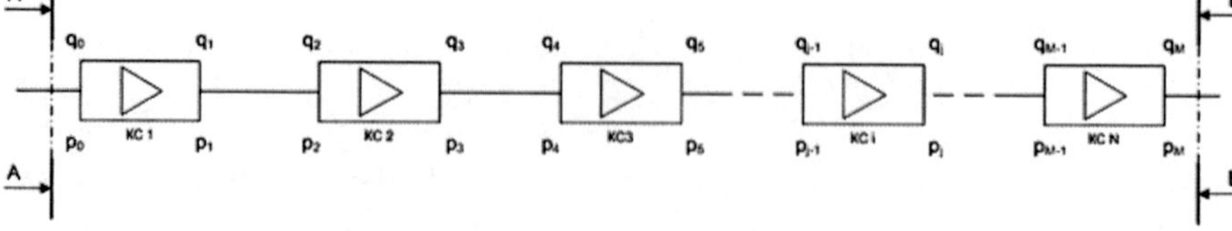

Fig. 5.13 Schematic design diagram of the gas pipeline

Let us consider the basic design scheme of the main gas pipeline shown in Fig. 5.13.

For the gas pipeline shown in Fig. 5.13, the boundary conditions will be p_0, p_M, q_0. The additive goal function for the scheme under consideration will have the following form: Equation Chap. 2 Section 1

$$\Phi = \varphi_1\left(p_0, \vec{u}_1\right) + \varphi_2\left(p_1, \vec{u}_2\right) + \ldots$$
$$\ldots + \varphi_i\left(p_{j-1}, \vec{u}_i\right) + \ldots + \varphi_N\left(p_{M-1}, \vec{u}_N\right) \tag{5.12}$$

Vector of controls $\vec{u}_i$ and the condition at the entrance to the compressor station p_{j-1} uniquely determine the resulting state p_j. And back, states p_{j-1} and p_j allow you to find management $\vec{u}_i$, although not necessarily unambiguously. We assume that for fixed p_{j-1} and p_j, we can find such management $\vec{u}_i$, in which the goal function would be minimal:

$$g_i\left(p_{j-1}, p_j\right) = \min_{p_j = p_j\left(p_{j-1}, \vec{u}_i\right), \vec{u}_i} \varphi_i\left(p_{j-1}, \vec{u}_i\right) \tag{5.13}$$

and minimization is carried out by those values $\vec{u}_i$, which lead to a given value p_j.

Function $g_j(p_{j-1}, p_j)$ represents the costs on the *j-th* element of the system, provided that the pressure in front of it is equal to p_{j-1}, and than p_j.

For a passive element due to the lack of controls for each p_{j-1}, there is only one valid value p_j, and the corresponding value of the function g_i is equal to zero.

For an active element, when $p_{j-1} = p_j$ и $g_i(p_{j-1}, p_j) = 0$, $q_j = q_{j-1}$ (the station is disconnected) or the pair p_{j-1} and p_j is technologically acceptable ($p_{j-1} < p_j$), and the corresponding constraints are met, then the function g_i, $i = 1..N$ is defined as follows:

- For gas turbine compressor station (C_g is the cost of 1000 m^3 of gas)

$$g_i\left(p_{j-1}, p_j\right) = C_g q_{m.z.}\left(p_{j-1}, p_j, q_{j-1}\right), q_j =$$
$$q_{j-1} - q_{m.z.}\left(p_{j-1}, p_j, q_{j-1}\right) \tag{5.14}$$

- For an electric compressor station (C_e is the cost per kWh)

$$g_i\left(p_{j-1}, p_j\right) = C_e E\left(p_{j-1}, p_j, q_{j-1}\right), q_j = q_{j-1} \tag{5.15}$$

If $p_j - p_{j-1} = \Delta p_j$ is the pressure change on the *j-th* element, then the following condition must be met:

$$\sum_{j=1}^{M} \Delta p_j = p_M - p_0 \tag{5.16}$$

The problem is reduced to finding a distribution that satisfies Eq. (5.11) and provides a minimum of the function:

$$R\left(p_1,\ldots,p_{M-1}\right)=g_1\left(p_0,p_1\right)+g_N\left(p_{M-1},p_M\right) \tag{5.17}$$

The principle of optimality can be directly applied to a multistep process. Let us introduce a sequence of functions $f_i(p_j)$ ($i = 1,\ldots,N$). Function f_i represents the total costs for the first j steps, provided that the trajectory $p_0, p_1, \ldots, p_j$ is optimal.

Naturally if $f_0(p_0) = 0$, then

$$f_1\left(p_1\right)=g_1\left(p_0,p_1\right) \tag{5.18}$$

For functions f_i ($i > 1$), a recurrence relation is obtained:

$$f_i\left(p_j\right)=\min_{p_{j-1}}\left[f_{i-1}\left(p_{j-1}\right)+g_i\left(p_{j-1},p_j\right)\right]\ \left(i=1\ldots\mathrm{N}\right) \tag{5.19}$$

Minimization is carried out by those values p_{j-1}, under which we will make the transition from the state p_{j-1} in the state p_j. Range of variation changes from step p_j to step $p_j \in [p_{*},p^{*}]$, where p^{*} is the maximal and p_{*} is the minimal value p. Values p^{*} and p_{*} will be determined by the appropriate set of ratios.

Let us put $g_j(p_{j-1},p_j) = \infty$, when, due to technological limitations, it is impossible to move from the state p_{j-1} in state p_j.

The numerical solution is obtained by discretization in the domain of phase coordinates p_j. Interval $[p_{*},p^{*}]$ is divided by a scale so that the values of the functions $f_i(p_j)$ at grid points p_j^l, they gave a fairly complete picture of the behavior f_i on the whole interval $[p_{*},p^{*}]$.

Thus, the general form of the equation by which the iterative process is constructed can be written as:

$$f_i\left(p_j^l\right)=\min_{\varsigma}\left[f_{i-1}\left(p_{j-1}^{\varsigma}\right)+g_i\left(p_{j-1}^{\varsigma},p_j^l\right)\right] \tag{5.20}$$

As an example of the implementation of this algorithm, consider a gas pipeline with a length of 963 km, a diameter of 1420 mm, which includes three threads and nine compressor stations (Fig. 5.14).

The optimization results are presented in Table 5.4

Thus, the analysis of the obtained results shows their adequacy, and the developed algorithms can be used for practical application in order to determine the emissions of pollutants during the operation of compressor stations on main gas pipelines. At the same time, the calculated operating modes of compressor stations make it possible to minimize the volumes of natural gas burned at the GPU at a given gas pipeline capacity. As noted in Samsonov et al. (2007b), and Rusakova et al. (2009), the ratio of fuel gas consumption to the amount of transported gas characterizes the

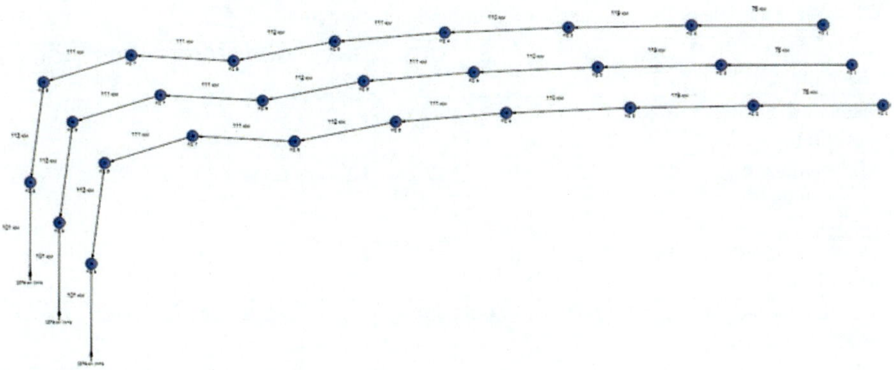

Fig. 5.14 Diagram of the section of the gas transmission system

efficiency of compressor stations (when the CS is operating according to the established technological scheme, this average indicator is estimated at 33 m^3/million m^3 *km). As a practical implementation of the proposed approach, an assessment of geoecological risks for the Yamal–West gas transmission system was carried out (see Chap. 3).

Optimization of geoecological risk management during the operation of compressor stations on main gas pipelines is associated with a reduction in emissions of harmful substances generated during the combustion of natural gas into the atmosphere. These substances, after photochemical and physico-chemical transformations in the atmosphere, can enter terrestrial and aquatic ecosystems in the CS impact zone, determining the background and impact effects on various components of these ecosystems. The permissible level of this impact can be calculated using the methodology of critical loads (Bashkin et al., 2002b). Exceeding the calculated values of critical loads (CL) leads to geoecological risks. In this case, it is necessary to reduce emissions of pollutants both through the use of technological operations and through the total reduction of energy consumption. Since these processes are expensive, it is necessary to estimate the magnitude of the prevented environmental damage (PED) in order to calculate their economic efficiency.

The following formula is used for the enlarged calculation of PED for ecosystem bioresources (soils, flora, and fauna) during the operation of GCS using the normative method:

$$\mathrm{PED} = \sum_{i=1}^{N} N_i^p \times \mathrm{K_p} \times R_\mathrm{d}, \tag{5.21}$$

where, PED is the monetary assessment of the amount of prevented damage to biological resources from measures to minimize energy consumption and the corresponding reduction of emissions of harmful substances during the operation of the GCS for the reporting period of time, thousand rubles/year

Table 5.4 Example of the optimization algorithm implementation

Name of GCS	GCS −1	GCS −2	GCS −3	GCS −4	GCS −5	GCS −6	GCS −7	GCS −8	GCS −9
Number of CS workshops	3	3	3	3	3	3	3	3	3
Max. number of gas pumping units (GPU)	15	12	10	12	12	12	12	12	10
Number of GPU under load	15	10	10	12	11	12	11	11	10
The number of GPU under load in each shop	/5/5/5/	/4/4/2/	/3/3/4/	/4/4/4/	/4/4/3/	/4/4/4/	/4/4/3/	/4/4/3/	/3/3/4/
Compression ratio	1.48	1.30	1.58	1.61	1.60	1.63	1.59	1.58	1.54
Outlet pressure CS, MPa	11.52	11.80	11.80	11.80	11.74	11.80	11.80	11.80	11.46
Inlet pressure CS, MPa	7.98	9.32	7.68	7.56	7.56	7.47	7.65	7.69	7.64
Gas consumption at the entrance to the CS, million m³/day	554.1	552.3	551.1	549.1	546.7	544.5	542.1	539.8	537.6
Temperature of the gas at the outlet of the CS, C°	12	−2	12	12	12	12	12	12	12
Gas temperature at the CS inlet, C°	−2	4	−19	−5	−5	−5	−4	−4	−4
Maximum available power of GPU, MW	264.0	211.2	275.0	330.0	330.0	330.0	330.0	330.0	275.0
Available capacity of working GPU, MW	264.0	176.0	275.0	330.0	302.5	330.0	302.5	302.5	275.0
Power on the coupling of the centrifugal pump, MW	262.6	172.4	267.4	306.9	301.7	313.6	295.4	290.7	274.2
Power load factor	0.99	0.98	0.97	0.93	1.00	0.95	0.98	0.96	1.00
Fuel gas consumption, million m³/day	1.79	1.18	2.03	2.35	2.27	2.39	2.24	2.21	2.07

N_i^p is the total number of species of biological resources (a complex of species from 1 to N) that can be lost because of inertial impact, pieces

R_d is the rate of damage to biological resources (the average value of the amount of taxes for each species from the analyzed complex of species in this territory), rubles.

The prevented damage to ecosystems in the GCS impact zone will be calculated as the difference between the magnitude of the predicted environmental damage caused without environmental protection measures (inertial scenario) and the residual environmental damage after their implementation.

5.3 Environmental Risk Insurance as a Risk Management Mechanism

The issues of insurance protection of environmental damage in recent years have become increasingly important due to the need to attract non-state resources in the field of environmental management to ensure effective industrial and environmental safety of various enterprises, prevention and elimination of accidents, and protection of human rights to a favorable environment. Thus, eco-insurance is included in the sphere of geoecological risk management. Detailed information on damage insurance related to environmental pollution, including in the gas industry, can be found in the works of Bashkin & Demidova (2006), Demidova et al. (2007), Lesnykh & Timofeeva (2004), Bashkin (2006a, b, 2009, 2017a, b), and Samsonov et al. (2007a).

Environmental insurance is a system of measures for the creation and use of a monetary (insurance) fund at the expense of contributions from its participants, from which economic damage to the environment and public health is compensated, as well as other monetary funds are paid in connection with the occurrence of certain events. The purpose of environmental insurance is to provide insurance for civil (property) liability for damage caused by an individual or legal entity as a result of sudden, unforeseen, and unintentional environmental pollution. This insurance can be mandatory (including state, i.e., municipal or federal) or voluntary.

In world practice, environmental insurance is a powerful financial mechanism for redistributing responsibility for compensation of damage from environmental pollution because of accidents, natural disasters, and catastrophes.

5.3.1 Insured Events in the Field of Nature Management and Environmental Protection

Economic harm to the environment and public health may arise from the separate or combined action of the following factors and situations, defined as “insured events”:

- Emergency and force majeure circumstances arising under unforeseen conditions, for example, the spill of fuel at various enterprises due to the destruction of product pipelines, food storage facilities, etc.; damage to treatment systems due to natural disasters and the breakthrough of drains; flooding of hazardous waste storage sites during floods, etc.

- Deferred damage is known damage to the environment and public health that occurs after a certain period of time due to the operation of this enterprise, for example, the development of chronic diseases with sublethal emissions of harmful substances (carcinogenic and non-carcinogenic diseases), gradual acidification of soils and natural waters with emissions of acid-forming compounds of sulfur and nitrogen even within the limits of permitted emissions, a decrease in the biodiversity of natural ecosystems in violation of their nutritional status due to the intake of nitrogen compounds emitted during the operation of the GCS, etc.
- Synergistic damage is damage produced by the joint action of those doses of contaminants that do not lead to harmful effects when considering effects of individual substances, for example, the carcinogenic effects of low doses of nitrogen oxidesNitrogen oxides and certain medical drugs, carcinogenic effects of the mixture of heavy metalsHeavy metals in small doses, etc.
- Lack of knowledge about the mechanisms of action of various emissions of pollutants for the period that leads to the need for the use of safety factors (i.e., the damage can be greater than 1 to more than 10 times).

Any enterprises (institutions, organizations) that are a legal entity and have production facilities on the territory of the Russian Federation can (and must) conclude an insurance contract with an insurance company to insure the environmental risk of third parties filing property claims against this enterprise (institution, organization) for compensation of damage from pollution of various environmental objects. In each specific case, when concluding a contract, a list of pollutants and causes of accidental pollution are specified, as well as damages which must be compensated. According to the concluded contract, the company undertakes to pay regularly the insurance company an insurance payment depending on the annual turnover of the company and the pre-agreed tariff rate.

5.3.2 Geoecological Insurance Scheme

In order to develop the regulatory and methodological framework necessary for the introduction of mandatory geoecological risk insurance for harm caused to public health and the environment by enterprises of various forms of ownership, a general scheme should be developed that allows implementing a mechanism for civil liability insurance for causing this harm. The scheme includes the development of the necessary regulatory and methodological documentation during several stages: (1) inventory of emissions of enterprises located in the territory of the administrative region under consideration; (2) approval of owners (operating organizations); (3) assessment of environmental risks; and (4) implementation of the insurance process itself.

Inventory of Emissions

The inventory of emissions of oil and gas enterprises on the territory of the Arctic administrative region carried out for the purpose of eco-insurance and the created database becomes an information base for all participants of the insurance market. The assessment of economic damages from emissions of atmospheric pollutants can be carried out based on specially developed models (see Chap. 3). This makes it possible to determine the tasks and activities of both potential policyholders (owners or tenants of enterprises) and insurers who acquire the opportunity to assess the capacity and potential of the market. This information is also required by the local administration and the territorial Department of Natural Resources, which can make the necessary decisions for each object, depending on the type of owner (operating organization), territorial status, and other conditions.

At the same time, the essential parameters of the operation of enterprises and their emissions into the environment identified in the process of inventory and database compilation can become the basis of a voluntary declaration of environmental safety. It allows all objects located in the district to be classified in the appropriate order and included in the regional register, which, in turn, gives the right to obtain a license and conduct production activities. Based on the results of the inventory, it is possible to compile a list of potential insurance claims that arise during the operation of the enterprise and cause harm to public health and the environment.

Approval of Owners (Operating Organizations)

The approval of the owners (operating organizations) of economic facilities in the process of implementing the mechanism of eco-insurance of civil liability for damage caused as a result of an accident caused by the owner or due to force majeure circumstances is an important component of this mechanism. This allows assessing the real situation with the ownership of enterprises to a particular government agency, private company, or other owner to determine the responsibility of the latter for ensuring the technological condition of the facility and to take the necessary measures to ensure the environmental and industrial safety of the facility. The owner is most interested in the financial security of enterprises, including at the expense of the insurance contract. In addition, the presence of the owner to a certain extent reduces the financial risks of insurance companies and thereby contributes to the reduction of insurance rates, which ultimately leads to the stability of insurance relationships in the field of environmental management.

Environmental Risk Assessment

The assessment of geoecological risk to public health and the state of the environment in the territory under consideration is carried out using the approaches discussed in Chap. 3. Based on the inventory of emissions of harmful substances into the atmosphere, the concentrations of various normalized compounds in the surface air layer in the district are calculated and compared with the adopted MAC and the morbidity of the resident population. The stability of terrestrial and aquatic ecosystems to the ingress of polluting compounds is calculated using the international methodology of critical loads. These critical loads are compared with the available intake of emitted chemical compounds into the atmosphere, surface and groundwater, and their storage in landfills of solid waste, which makes it possible to assess the presence of excess critical loads. A comparative analysis of the concentration fields and the values of exceedances of critical loads allows us to characterize the environmental situation in various parts of the region, both in urban and in rural areas, and in various uninhabited spaces, also exposed to emissions from enterprises.

The application of such a realistic and scientifically based approach will allow launching a mechanism for eco-insurance of enterprises, taking into account the capabilities of all participants in insurance relations. Tariff rates for environmental insurance are set in percent of the insured amount (liability limit), in percent of the amount of penalty payments for the previous period equal to the period of validity of the insurance contract, and in percent of annual production. Attempts to establish speculative insurance amounts, premiums, and tariffs discredit the process of introducing insurance methods in the field of environmental management. In fact, having no information in recent years about the magnitude of environmental risk, accidents that have occurred, and the funds spent on eliminating their consequences, it is impossible to statistically reliably assess the damage to the environment and public health from the activities of enterprises in one or another part of the Arctic region with different resistance to pollutants. Without this, it is impossible to calculate the probability of an insured event and the damage from it, and, accordingly, the amount of the insured sum. Therefore, at this stage, as at others, a factual approach is necessary.

Implementation of the Eco-insurance Process

The implementation of the actual eco-insurance process should begin with an assessment of the value of the enterprise, its products, and the amount of emissions into the environment, which gives an idea of its financial situation, according to which a realistic structure of the insurance tariff can be developed. At the same time, together with the owners and local authorities, various sources of financing for guarantees of environmental safety of enterprises are being considered. The implementation of the eco-insurance process takes place through the registration of relevant insurance documents, the standard forms of which are approved (coordinated) with the supervisory authorities.

In case of damage to public health and the environment caused by the insured enterprise as a result of an accident and/or other insured events that led to pollution of various components of terrestrial ecosystems, air, surface and groundwater, the insurance company is obliged to pay insurance compensation, including the following:

- The amount of losses associated with the deterioration of living conditions of the population and the environment
- Expenses necessary to save the life and property of third parties who have suffered damage as a result of the accident
- The costs of cleaning up polluted environmental objects (water, air, terrestrial ecosystems) and bringing them into a state corresponding to the accepted standards
- Related to the preliminary investigation, the conduct of trials and other costs of settling any claims brought against the company

In addition to the task of compensating losses, environmental insurance is designed to stimulate the implementation of preventive environmental measures. At the same time, the amount of the insurance tariff is differentiated depending on the type of production, the risk of the production process, and the implementation of preventive measures at each specific enterprise. In case of trouble-free operation, insurance payments are reduced annually. Also in agreement with the insurance company, the company has the opportunity to spend part of the insurance payments on targeted financing of its own environmental programs and measures to improve environmental safety.

The introduction of compulsory insurance of civil liability for harm caused to public health and the environment as a result of the activities of enterprises will require an institutional mechanism that allows for rational regulation of funds accumulated in insurance funds. These funds can be used both to develop preventive measures to reduce environmental risk and to eliminate the environmental consequences of accidental emissions and discharges.

At the first stages of the development of the environmental insurance system in a certain administrative region, it is primarily advisable to carry it out at enterprises with increased danger of the production process and high absolute and specific emissions of waste into the environment. In addition, the conclusion of an insurance contract should be mandatory when registering new enterprises and/or reapproving permissible regulatory emissions.

In general, the introduction of eco-insurance is carried out to solve priority tasks related to ensuring the safety of enterprises and aimed at the development of the territory of the Arctic region. Environmental insurance does not replace environmental payments (for known, permissible emissions damage or to pay for environmental management), but is additional to cover the costs of environmental rehabilitation and restoration of public health due to the above combination of various factors.

5.4 Managing Geopolitical Risks in the Arctic Regions

As noted earlier, there are positive examples of managing geopolitical risks in the Arctic region based on international law, and clear rules and regulations arising from its provisions. It can be stated that this is the cornerstone reason for the peaceful settlement of disputes in the region. Moreover, all major non-Arctic states (Japan, China, India, Malaysia, etc.) have declared their commitment to the geopolitical practice of interaction between countries adopted in this region. At the same time, within the framework of the topic of geopolitical risk management, there are opportunities for further improvement of these rules. It is also necessary to take into account the increased interest of non-Arctic countries in participating in a number of oil and gas projects, for example, Russian Arctic LNG projects. In accordance with the proposals made earlier (Collins, 2017a, b), a number of practical steps in this direction can be considered:

- Expanding the scope of the Arctic Council to include security, trade, and immigration issues. Non-Arctic countries can be accepted as full or associate members of this organization, provided they meet a number of criteria (for example, participation in regional research activities, entry into Arctic oil and gas projects, etc.).
- The establishment of an International Arctic Boundary Delimitation Commission to settle disputes between Arctic States, as well as the adoption of a possible Arctic Treaty, which could become a larger structure and forum for both Arctic States and non-Arctic countries.
- Implementation of practical steps to assess the likelihood of transformation of geopolitical risks into opportunities and/or threats, which is one of the priorities of the Arctic oil and gas industry facilities.
- The need to prioritize the management of the following key geopolitical risks: (1) ensuring access to sufficient reserves of hydrocarbon raw materials of the Arctic by various states, obtaining control rights over its natural resources; (2) uncertainty of the legal status of the Arctic region; and (3) geoecological risks as one of the priorities of attention to Russia's actions in the Arctic.

5.5 Conclusions

Thus, the considered examples, affecting the three basic directions in the work of oil and gas companies, in particular, gas producing and gas transportation companies, such as technological, geoecological, and environmental, reveal the need to revise a number of provisions on the assessment of various risks of Arctic and subarctic oil and gas projects. In essence, the solution of risk management tasks in these three directions, starting with their formulation, must necessarily be object-oriented, i.e., aimed at preventing a potential catastrophe. At the same time, the answer to the question that arises, "is it possible to prevent disasters at all?" boils down to the following arguments, which have both a methodological and technological nature. First, it should be emphasized that this is necessary, although it is not possible in all cases. Previously, the most common methods of risk management, including prevention of the potential development of disasters, were, in particular, increasing the safety margin, removing work from the bifurcation points of technological processes, overspending of an inhibitor to prevent hydrate formation, etc. Such methods are sometimes used now, but entail large additional costs. At the same time, it is possible to work with the threat of a catastrophe, because its development requires a certain amount of time (for example, it takes one to two years to manage the risks of pollution of Arctic soils, and about 100 ms for a gas turbine engine). This time is usually called the event horizon. In this case, the reaction to the emerging situation should follow no later than before half of the time allotted in a particular situation expires, and all technological solutions must be determined by the duration of the transients of this particular technology, regardless of its nature. And most importantly, a technological assessment of the event horizon and an orientation toward its assessment is necessary.

This approach allows you to manage the risks of the projects under consideration and achieve their sustainable development. Taking into account the existing realities of the Arctic region, geopolitical and geo-ecological risks will be among the priorities, the management of which will make the greatest contribution to the development of oil and gas projects.

Conclusions

The monograph is devoted to the issues of geoecological risks (GER) and geopolitical risks (GPR) due to the hydrocarbon development of the Arctic. Although the main focus is on the Russian part of this region, many assumptions and conclusions are quite applicable to the entire Arctic basin.

The authors reveal the ethical and philosophical foundations of understanding environmental equity and justice in relation to the development of Arctic natural resources and consider various approaches to the moral aspects of making global management decisions in the field of environmental protection and environmental safety.

At the same time, special attention is paid to the analysis of modern geopolitical and geostrategic challenges to the development of the Arctic region. This monograph identifies the key geopolitical factors affecting the sustainable development of the Arctic and analyzes the similarities and differences in the geostrategic positions of the Arctic Five states. One of the most important factors of the twenty-first century that determine the alignment and interaction of various geopolitical forces is the struggle for resources. In this regard, an increased GPR in the Arctic, related to its resource potential, is inevitable. For objects of the oil and gas industry, geopolitical as well as geoecological risks can be transformed into opposite factors in the form of additional opportunities or threats, which the authors identify in detail for each type of risk. The authors focus on such positions of the GPR, which are related to ensuring access and obtaining control rights over the Arctic's hydrocarbon resources from different countries, the uncertainty of the legal status of the Arctic region, and the use of GER as manipulative tools for attention to Russia's actions in the Arctic.

The main aspect of the monograph is geoecology in the Arctic. The authors reveal the need to consider geopolitical challenges in the process of GER analysis for oil and gas development in the Arctic region. This is due to the intersection of the strategic interests for a number of States and their focus on proving Russia's

V. N. Bashkin, O. P. Trubitsina, *Geoecological and Geopolitical Risks for the Oil and Gas Industry in the Arctic*, Environmental Pollution 29,
https://doi.org/10.1007/978-3-030-95910-4

inability to ensure environmental safety in the development of Arctic deposits. The monograph analyzes in detail each type of GER with an emphasis on their transformation into additional opportunities and threats. The authors propose a model of GER analysis, which is based on critical loads (CL) of pollutants and includes two stages: (1) the stage of GER quantitative assessment, which allows us to calculate not only the magnitude of predicted changes in the state of Arctic ecosystems, but also the probability of their occurrence and (2) the stage of GER management, taking into account geopolitical factors, which involves a qualitative expert assessment, which is a management decision-making procedure for achieving acceptable levels of total GER.

This monograph analyzes the environmental rating as an indicator of the GER management process of Russian oil and gas companies operating in the Arctic. The authors demonstrate the algorithm of the GER management model processes and reveal the necessity of using environmental ratings for the oil and gas industry. This book also focuses on the issues of environmental rating as an indicator of geoecological risk management of Russian oil and gas companies in the Arctic. The authors discuss the need of using environmental rating for oil and gas industry in Russia. Primarily, specialized attention is given to the issues of rating results of environmental responsibility of the relevant companies in Russia that was first held in 2014 by the cooperative initiative by CREON Group and WWF Russia with participation of National Rating Agency. The authors have selected and analyzed from all Russian oil and gas companies only those who are operating in the Arctic region.

The rating's results show the "top three" companies among Russian oil and gas companies operating in the Arctic – Zarubezhneft, Lukoil, and Gazprom (Sakhalin Energy). There are also other companies in rating's sectors. For instance, Zarubezhneft and LUKOIL share the best position in terms of environmental management (the 1st and the 2nd positions). The best companies in "environmental impact" sector are Gazprom, Zarubezhneft, and LUKOIL, whereas the leaders in "disclosure/transparency" sector are Sakhalin Energy (Sakhalin-2) and Zarubezhneft (the 1st and the 2nd positions).

In whole, the environmental rating of the Russian oil and gas companies, operating in the Arctic, may create a new mechanism of identification of their CSR level. Particularly, analysis of a number of criteria of environmental management, environmental impact, and information disclosure or transparency helps correlate the process and results of economic activity of oil and gas companies in the Arctic with the requirements and expectations of society. Along with it, the desire to increase these indicators contributes to investment attractiveness of oil and gas companies about their environmental responsibility. It helps to inform foreign investors; as a result, it may lead to the rise of Foreign Direct Investments (FDI) in Russia and decline of the GER in conditions of Arctic's exploitation accordingly.

It is important to emphasize that this monograph examines the features of the analysis and risk management of oil and gas projects in the Arctic region. It is shown that in the course of risk management of oil and gas projects in the Arctic region, it is necessary to take into account the geopolitical, climatic, ethical, institutional, geoecological, technological, financial, and other features of the implementation of projects. Accordingly, the diversity of these factors imposes increased requirements on them related to the analysis of risks of various nature and their interrelation. The algorithms for considering each type of risk are presented. The methods of managing geoecological risks on complex main gas pipelines passing through the territory of the coastal Arctic regions are analyzed separately. Mathematical models are presented to minimize the costs of gas transportation through complex gas transmission networks. The necessity of taking into account, along with technical and economic indicators, also geoecological and ecological components of the functionals is shown. It is demonstrated that in the economic aspect, the costs of reducing (geo)ecological risks are significantly lower than the total energy costs. Nevertheless, the optimization of gas flows through gas transmission networks, taking into account the economic assessment of the gas transport system (GTS) impact on the environment, will make it possible to develop solutions for managing geoecological risks. Next, such an approach to geoecological risk management as insurance is considered. The issues of insurance protection of environmental damage in recent years have become increasingly important due to the need to attract non-state resources in the field of environmental management to ensure effective industrial and environmental safety of various enterprises, prevention and elimination of accidents, and protection of human rights to a favorable environment. This is of the special attention in the Arctic region.

In the end, the consideration of geoecological and geopolitical aspects of the development of Arctic resources on the example of Russian oil and gas companies makes it possible to find a consensus on the scale of the entire Arctic region, taking into account the interests of all other national and international companies involved in this process.

References

Abramov, A. N., Aniskovich, M. F., & Yudina, M. P. (1992). Structure and oil and gas potential of the Yurobcheno-Takhoma zone according to the data of complex interpretation of geological and geophysical data. *Applied Geophysics, 127*, 23–35.

Akaah, I. P. (1997). Influence of deontological and teleological factors on research ethics evaluation. *Journal of Business Research, 39*, 71–80.

Allayarov, R. A., & Shubin, S. I. (2017). Strategic interests of Denmark in the Arctic. Greenland is the Main factor in preserving the Arctic status. *Problems and Prospects Student Science Issues, 15*, 198–202.

Alekseeva, M. N., Svarovskaya, L. I., & Yashchenko, I. G. (2019). Risks of pollution of Arctic territories by oil and oil products. *AIP Conference Proceedings, 2167*(1), 020009. https://doi.org/10.1063/1.5131876 (International conference on advanced materials with hierarchical structure for new technologies and reliable structures).

Arabsky, A. K., & Bashkin, V. N. (2015). Identification, challenges and ways of addressing (geo)ecological risks at the regional scale. *Open Ecology Journal, 8*, 86–91. https://doi.org/10.2174/1874213001508010086

Arabsky, A. K., Bashkin, V. N., & Galiulin, R. V. (2015). Ecological risk challenge at Gazprom Dobycha Yamburg LLC. *Open Ecology Journal, 8*, 78–85. https://doi.org/10.2174/1874213001508010078

Arabsky, A. K., Nikolaev, D. S., & Bashkin, V. N. (2016). Identification of geoecological hazard in connection with the emission of pollutants in the gas fields of the Far North. *Environmental Yearbook, 9*, 58–63.

Aristotle. (1925). *The Nicomachean ethics of Aristotle* (p. 133). Oxford University Press.

Arctic standards recommendations on oil spill prevention, response, and safety in the U.S. Arctic Ocean. (2013, Sept). A report from the PEW charitable trust (p. 139).

Arno, O. B., Arabsky, A. K., & Bashkin, V. N. (2021). Innovative development of Gazprom Dobycha Yamburg LLC. Similarities and differences in methods of preventing potential disasters. *Gas Industry, Special Issue, 1*(184), 26–34.

Arnold, C. A. (1998). Planning Milagros: Environmental justice and land use regulation. *Denver University Law Review, 76*(1), 1–153.

Ausubel, J. H., & Victor, D. G. (1998). Verification of international environmental agreements. *Annual Review of Energy and the Environment, 17*, 1–43.

Baalsrud, K., & Magnusson, J. (1990). *Eutrophication in Oslofiord in 1989*. Noverdrapport. SFT/NIVA-report 427/90. 43 pp.

V. N. Bashkin, O. P. Trubitsina, *Geoecological and Geopolitical Risks for the Oil and Gas Industry in the Arctic*, Environmental Pollution 29,
https://doi.org/10.1007/978-3-030-95910-4

Baklanov, P. Y., Moshkov, A. V., & Romanov, M. T. (2015). Geographical and geopolitical factors and directions of the long-term development of the Arctic zone of Russia. *Vestnik of the Far East Branch of the Russian Academy of Sciences, 2*, 5–15.

Baden, B. M., & Coursey, D. L. (2002). The locality of wastes sites within the city of Chicago: A demographic, social and economic analysis. *Resource and Energy Economics, 24*, 53–93.

Banuri, T., Goran-Maler, K., Grubb, M., Jacobson, H. K., & Yamin, F. (1996). Equity and social considerations'. In J. P. Bruce, H. Lee, & E. F. Haites (Eds.), *Climate change 1995: Economic and social dimensions of climate change*. Contribution of Working Group III to the Second Assessment Report of the Intergovernmental Panel on Climate Change. (pp. 83–124). Cambridge University Press.

Baron, J. (1994). Nonconsequentialist decisions. *Behavioral and Brain Sciences, 17*, 1–42.

Bashkin V. N. (2002). *Modern biogeochemistry*. Kluwer Academic Publishers. 591 pp.

Bashkin, V. N. (2006a). *Modern biogeochemistry: Environmental risk assessment* (2nd ed., p. 444). Springer Publishers.

Bashkin, V. N. (2006b). *Environmental risks: Assessment, management, insurance* (p. 360). Higher School.

Bashkin V. N. (2009). *Modern biogeochemistry: environmental risk assessment*, 2d Ed. Chinese translation, CIP. 268 pp.

Bashkin, V. N. (2014). *Biogeochemistry of polar ecosystems in the zones of influence of the gas industry*. Gazprom-VNIIGAZ. 302 pp.

Bashkin, V. (Ed.). (2016). *Biogeochemical technologies for managing pollution in polar ecosystems*. Environmental pollution (Vol. 26, p. 219). Springer. https://doi.org/10.1007/978-3-319-41805-6

Bashkin, V. N. (2017a). Biogeochemical cycles in tundra ecosystems in areas impacted by gas industry facilities. *Geochemistry International, 55*(10), 946–956. https://doi.org/10.1134/S0016702917100020

Bashkin, V. (Ed.). (2017b). *Ecological and biogeochemical cycling in impacted polar ecosystems* (p. 308). NOVA Publishers.

Bashkin, V. (2018). Carbon biogeochemical cycle and consequences of climate changes. In B. Fath (Ed.), *Encyclopedia of ecology* (2nd ed.). Elsevier, ISBN: 9780444637680.

Bashkin, V. N., Park, S.-U., & Choi, M. S. (2002a). Regional budget of nitrogen in North-East Asia. *Biogeochemistry, 57*(1), 387–403.

Bashkin, V. N., Kazak, A. S., Snakin, V. V., Priputina, I. V., Khrisanov, R. V., & Kochurov, B. I. (2002b). *Stability of ecosystems to the emissions of main gas pipelines* (p. 231). Universum.

Bashkin, V. N., Erdman, L. E., Kozlov, M. Y., & Priputina, I. V. (1995). Input of atmospheric precipitation nitrogen into the Baltic Sea from its watershed. *Water, Air, and Soil Pollution, 85*, 871–876.

Bashkin, V. N., & Priputina, I. V. (2010). *Management of environmental risks in the emission of pollutants* (p. 189). Gazprom-VNIIGAZ.

Bashkin, V. N., Arno, A. B., Arabsky, A. K., Barsukov, P. A., Priputina, I. V., & Galiulin, R. V. (2013). *Retrospective and forecast of the geoecological situation at the gas condensate fields of the Far North* (p. 280). Gazprom-VNIIGAZ.

Bashkin, V. N., & Galiulin, R. V. (2019). *Geoecological risk management in polar areas*. Environmental Pollution, 28 (p. 155). Springer. https://doi.org/10.1007/978-3-030-04441-1

Bashkin, V. N., & Demidova, O. A. (2006). Insurance of environmental risks. *Oil of Russia, 1*, 25–34.

Bashkin, V. N., & Galiulin, R. V. (2015). Ecological management of (geo)ecological risks in gas industry. *Open Ecology Journal, 8*, 73–77. https://doi.org/10.2174/1874213001508010073

Bashkin, V. N., & Demidova, O. A. (2008). *Biogeochemical approaches to environmental risk assessment*. Encyclopedia of Ecology (pp. 378–386). Elsevier.

Bashkin, V. N., & Lesnych, V. V. (2015). Systemic risks management and ecologically oriented gas industry development strategy. *Open Ecology Journal, 8*, 61–64. https://doi.org/10.2174/1874213001508010061

Bashkin, V. N., & Kazak, A. S. (2015). Geo-ecological risks and implementation of master scheme for gas industry development. *Open Ecology Journal, 8*, 69–72. https://doi.org/10.2174/1874213001508010069

Bashkin, V. N., & Priputina, I. V. (2015). (Geo) ecological risk assessment in gas industry development scenarios. *Open Ecology Journal, 8*, 65–68. https://doi.org/10.2174/1874213001508010065

Bashkin, V. N., Trubitsina, O. P., & Priputina, I. V. (2016) Evaluation of geo-environmental risks in the impacted zones of oil and gas industry in the Russian Arctic. In V. Bashkin (Ed.), *Biogeochemical technologies for managing environmental pollution in polar ecosystems* (pp. 109–116).

Bashkin, V. N., & Trubitsina, O. P. (2018). Geoenvironmental and geopolitical risks as the activities; challenges for the oil and gas industry in the Arctic. *Neftegaz.RU, 6*, 82–86.

Bass, J. (1998). Evaluating environmental justice under the National Environmental Policy Act. *Environmental Impact Assessment Review, 18*, 83–92.

Been, V. (1993). What's fairness got to do with it? Environmental justice and the siting of locally undesirable land uses. *Cornell Law Review, 78*, 1001–1085.

Been, V., & Gupta, F. (1997). Coming to the nuisance or going to the barrios? A longitudinal analysis of environmental justice claims. *Ecology Law Quarterly, 24*, 1–56.

Belov, N. P., & Baranovskaya, A. V. (1969). *Soils of the Murmansk region* (p. 147). Nauka.

Bhaskar, V. (1995). Distributive justice and the control of global warming. In V. Bhaskar & G. Glyn (Eds.), *The north the south and the environment: Ecological constraints and the global economy* (pp. 102–117). Earthscan Publications Ltd.

Bigliani, R. (2013, May). *Reducing risk in oil and gas operations. White paper*. Sponsored by: EMC. IDC energy insights (p. 15).

Bloom, E. T. (1999). The establishment of the Arctic council. *The American Journal of International Law, 93*(3), 714, 718, 720, 721–722.

Bobylev, N. G., Gadal, S., Konovalova, M. O., Sergunin, A. A., Tronin, A. A., & Tyunkyunen, V. P. (2020). Regional ranking of the Arctic zone of the Russian Federation on the basis of the environmental security index. *Sever & Rynok: formirovanie ekonomicheskogo poryadka, 3*, 17–40.

Bogolyubov, S. A., & Krasnova, I. O. (2018). Law and protection of the nature of the Russian Arctic. *Actual Problems of the Russian Law, 6*(91), 178–190.

Brashear, T. G., Brooks, C. M., & Boles, J. S. (2002). Distributive and procedural justice in a sales force context scale development and validation. *Journal of Business Research, 5733*, 1–8.

Bratyushchenko, S. V., & Seliverstov, V. E. (2007). Corporate social responsibility as an institution of public-private partnership and efficient regional economic and social policy. *Region: Economics and . Institute of Economics and Industrial Engineering of Siberian Branch of RAS, 4*, 189–206.

Buchanan, A. & Mathieu, D. (1986). Philocophy and justice. In R. L. Cohen (Ed.), *Justice: Views from the social sciences* (pp. 11–45). Plenum Press.

Bullard, R. (1999). Environmental justice challenges at home and abroad. In N. Low (Ed.), *Global ethics and environment* (pp. 33–46). Routledge.

Byrne, J., Wang, Y., Lee, H., & Kim, J. (1998). An equity- and sustainability-based policy response to global climate change. *Energy Policy, 26*(4), 335–343.

Carter, M. W. (2002). Advancing an ethical framework for long-term care. *Journal of Aging Studies, 16*, 57–71.

Camacho, D. E. (1998). *Environmental injustices, political struggles: Race, class, and the environment* (p. 232). Duke University Press.

CCE Report. (2005). *European critical loads and dynamic modelling. CCE status report 2005*. 171 p.

Chase, S. (2013, Dec 4). Harper orders new draft of Arctic seabed claim to include North pole. *The globe and mail*.

Chernyaev, V. D., Yakovlev, Y. I., Kazak, A. S., & Soshchenko, A. Y. (1991). *Pipeline transportation of hydrocarbon new materials* (p. 343). VNIIOEOG.

Collins, J. F. (2017a). The Arctic in the period of geopolitical changes: Assessment and recommendations. *Russia in Global Politics., 06*(10), 17.

Collins, J. F. (2017b). *The Arctic in an age of geopolitical change: Assessment and recommendations.* Valdai Papers. URL: https://eng.globalaffairs.ru/articles/the-arctic-in-an-age-of-geopolitical-change-assessment-and-recommendations/. Assessed by 12 Oct 2021.

Commodity Cycles: Oil prices will fall in 2020. (2013). https://www.interfax.ru/business/327893. Accessed 04 Feb 2021.

Dasgupta, P. (1990). Well-being and the extent of its realization in poor countries. *Economic Journal, 100*, 1–32.

Decree of the President of the Russian Federation of October 26, 2020. No. 645 "*On the Strategy for the Development of the Arctic Zone of the Russian Federation and Ensuring National Security for the Period up to 2035*".

Demidova, O. A., Bashkin, V. N., Yurkin, E. A., Kotova, M. V., & Priputina, I. V. (2007). Assessment of ecosystem risk with an increase in aluminum production at the Timan-Pechora deposit. *Issues of Risk Analysis, 2*, 140–151.

Dentener, F., Drevet, J., & Lamarque, J. F. (2006). Nitrogen and sulfur deposition on regional and global scale: A multimodel evaluation. *Global Biogeochemical Cycles, 20*, GB4003.

Diamantaras, D., & Thomson, W. (1990). A refinement and extension of the no-envy concept. *Economic Letters, 33*, 217–222.

Diagnostic analysis of the environmental status of the Russian Arctic (Advanced Summary). (2011). Editor-in-Chief: B.A. Morgunov, Moscow, Scientific World (p. 172).

Dushkova, D. O., & Evseev, A. V. (2011). Analyses of technogenic impact on geosystems of the European Russian North. *Arctic and North, 4*(November), 1–27.

Douma, W. T. (2010). The EBRD and Russia: Stimulating European principles for the environment. In W. T. Douma & F. M. Mucklow (Eds.), *Environmental finance and responsible business in Russia: Legal and practical trends* (pp. 169–188). T.M.C. Asser Press.

Dworkin, R. (1993). Integrity. In M. Fisk (Ed.), *Justice* (p. 140). Humanities Press.

Eglin, R. (2001). Keeping the «T» in the WTO: Where to next on environment and labour standarts? *The North American Journal of Economics and Finance, 12*(2), 173–191.

Environmental Management in Oil and Gas Exploration and Production. (1997). Joint E&P Forum/UNEP Technical Publication. An overview of issues and management approaches. Oxford, United Kingdom (p. 68).

Environmental Standards for Operations of Oil and Gas Companies Acting in Russia, on its Continental Shelf, and within its Exclusive Economic Zone developed by Russian Non-governmental Nature Conservation Organizations. (2005). (p. 8).

Exner-Pirot, H. (2016, Oct 14). Why governance of the North needs to go beyond the Arctic Council? *Open Canada*. Org. URL: https://www.opencanada.org/features/why-governance-north-needs-go-beyond-arctic-council/. Accessed 12 Oct 2021.

Fairman, R., Mead, C. D., & Williams, P. W. (1999). *Environmental risk assessment: Approaches, experiences and information sources* (p. 219). European Environment Agency.

Fedorov, E. S. (2020). Northern Sea route: The ice is melting — The tension is growing. URL: https://topwar.ru/178138-severnyj-morskoj-put-ldy-tajut-naprjazhennost-rastet.html. Accessed 11 Oct 2021.

Foster, S. (1993). Race(ial) matters – The quest for environmental justice. *Ecology Law Quarterly, 20*(4), 721–753.

Freeman, R. E. (2004). Stakeholder theory of the modern corporation. In T. Donaldson, P. H. Werhane, & M. Cording (Eds.), *Ethical issues in business* (pp. 38–49). Prentice Hall.

Garmo, O. A., Skjelkvale, B. L., de Wit, H., Colombo, L., Curtis, C., Folster, J., Hoffman, A., Hruska, J., et al. (2014). Trends in surface water acidified areas in Europe and North America from 1990 to 2008. *Water, Air, and Soil Pollution, 225*, 1–14.

Gershelis, E. V., Kashapov, R. S., Ruban, A. S., Oberemok, I. A., Leonov, A. A., Chernykh, D. V., Dudarev, O. V., & Semiletov, I. P. (2020). Identifying sources of organic carbon in surface sediments of Laptev Sea shelf using a rock-Eval approach. *Bulletin of the Tomsk Polytechnic University*. Geo Assets Engineering, *331*(8), 189–198.

Global and Russian Energy Outlook to 2040. (2014). Research advisor: A. A. Makarov, Project Manager: T. A. Mitrova, Professor L. M. Grigoryev. ERI RAS, ACRF (p. 175).

Golubeva, N. A. (2007). The pollution of the Arctic by toxic heavy metals. In G. G. Matishov (Ed.), *Biology and oceanography of the Northern sea route: Barents and Kara seas* (pp. 173–189). Nauka.

Gorobets, A. (2019). *New US Arctic Policy*. URL: https://icds.ee/ru/novaja-arkticheskaja-politika-ssha/. Accessed 10 Oct 2021.

Gramberg, I. S., Kulakov, Y. N., Pogrebitsky, Y. E., & Sorokov, D. S. (1983). *Arctic oil and gas superbassin.* Proc. X World Petroleum Congress. London (pp. 93–99).

Gramberg, I. S., & Pogrebnitsky, Y. E. (Eds.). (1984). *Geological structure and regularities of mineral placement. Vol. 9. Seas of the Soviet Arctic* (p. 280). Nedra.

Gramberg, I. S., & Laverov, N. P. (Eds.). (2000). *The Arctic on the threshold of the third millennium (resource potential and environmental problems)* (p. 247). Nauka.

Grinko, A. A., Goncharov, I. V., Shakhova, N. E., Gustafsson, O., Oblasov, N. V., Romankevich, E. A., Zarubin, A. G., Kashapov, R. S., Chernykh, D. V., Gershelis, E. V., Dudarev, O. V., Mazurov, A. K., & Semiletov, I. P. (2020). Sediment organic matter in areas of intense methane re-lease in the Laptev Sea: Characteristics of molecular composition. *Russian Geology and Geophysics, 61*(4), 560–585.

Hall, J., Ullett, J., Heywood, L., et al. (2003). *Status of UK critical loads: Critical loads methods, data and maps.* http://critloads.ceh.ac.uk

Harding, L. (2010, Sept 15). *Russia and Norway to resolve Arctic border dispute*. The Guardian. URL: https://www.theguardian.com/world/2010/sep/15/russia-norway-arctic-border-dispute

Howarth, R. W. (1996). *Nitrogen cycling in the North Atlantic Ocean and its watersheds* (p. 304). Kluwer Academic Publishers.

Hessen, D., Vadstein, O., & Magnusson, J. (1992). Nitrogen to marine areas, on the application of a critical load concept. *Critical Loads for Nitrogen*. Nord, *41*, 161–237.

Huebert, R., Exner-Pirot, H., Lajuenesse, A., & Gulledge, J. (2012). *Climate change & international security: The Arctic as a Bellwether* (pp. 1–3). Center for Climate and Energy Solutions.

ICJ Reports. (1985). *Case concerning the continental shelf – Malta/Libya*. The Hague.

Ikeme, J. (2003). Equity, environmental justice and sustainability; Incomplete approaches in climate change politics. *Global Environmental Change, 13*, 197.

Improving safety and productivity in oil and gas operations. (2014). White Paper. Motorola solutions. (p. 14). URL: https://www.motorolasolutions.com/content/dam/msi/docs/en-xu/oil-and-gas/Oil_and_Gas_Safety_and_Productivity_White_Paper.pdf. Accessed 21 Oct 2021.

Kaminsky, V. D. (2009). *The deep structure of the Central Arctic basin (in connection with the justification of the outer boundary of the continental shelf of the Russian Federation and the assessment of hydrocarbon resources).* Abstract of the dissertation for the degree of Doctor of Geological and mineralogical Sciences. St. Petersburg: (p. 48).

Kant, I. (1966). *Fundamentals of metaphysics of morals* (Vol. 2, p. 72). Nauka.

Kazak, A. S., Yakovlev, Y. I., & Kudryavtseva, T. A. (1985). *The system analysis of oil and gas transport mains*. Textbook (p. 76). MINKhGP.

Kazak, A. S., Sedov, V. I., Orechov, I. V., & Yakovlev, Y. I. (1989). *Operational control of main gas pipelines* (p. 289). Nedra.

Kazak, A. S., & Bashkin, V. N. (2015a). Cost-effectiveness of gas industry development strategy. *Open Ecology Journal, 8*, 56–60. https://doi.org/10.2174/1874213001508010056

Kazak, A. S., & Bashkin, V. N. (2015b). Methodology for gas industry sustainable development strategy based on the ecological imperatives. *Open Ecology Journal, 8*, 44–53. https://doi.org/10.2174/1874213001508010044

Knizhnikov, A. Y., Pakhalov, A. M., Sipailova, Y., & Rodionov, K. (2015). *Environmental responsibility rating of oil and gas companies in Russia* (p. 22).

Kolesnikov, E. Y. (2008). The practice of developing safety data sheets. *Issues of Risk Analysis, 4*(2), 106–128.

Kolosov, E. S. (Ed.). (1997). *Current international law: in 3 vols*. Moscow: Moscow state institute of international relationships, Moscow independent institute of international law, 688.

Komleva, N. A. (2011). Arctic strategy of the subarctic states: General and especial. *Arctic and North, 2*, 19–25.

Konow, J. (2001). Fair and square: The four sides of distributive justice. *Journal of Economic Behaviour & Organization, 46*, 137–164.

Konyshev, V. N., & Sergunin, A. A. (2011). *The Arctic in international politics: Cooperation or rivalry?* (p. 194). Russian Institute for Strategic Studies.

Konyshev, V. N., Sergunin, A. A., & Subbotin, S. V. (2017). State priority – Sustainable development of the Russian Arctic. *National Interests: Priorities and Security, 3*(348), 416–430.

Kovalsky, V. V. (1974). *Geochemical ecology*. Nauka, 300 pp.

Kuersten, A. (2016). The Arctic five versus the Arctic council. In L. Heininen, H. Exner-Pirot, & J. Plouffe (Eds.), *The Arctic yearbook: 20 years of regional cooperation and policy-shaping* (pp. 389–395). Northern Research Forum/University of the Arctic.

Kumpula, T., Pajunen, A., Kaarlejärvi, E. M., Forbes, B. C., & Stammler, F. (2011). Land use and land cover change in Arctic Russia: Ecological and social implications of industrial development. *Global Environmental Change, 21*(2), 550–562.

Lambert, T., & Boerner, C. (1997). Environmental inequity: Economic causes, economic solutions. *Yale Journal on Regulations, 14*, 195–234.

Lerberg, S. B., Holland, A. F., & Sanger, D. M. (2000). Responses of tidal creek macrobenthic communities to the effects of watershed development. *Estuaries, 23*, 838–853.

Lesnykh, V. V., & Timofeeva, T. B. (2004). Catastrophic bonds as a mechanism for financing the risks of catastrophic situations. *Issues of Risk Analysis, 1*(1), 154–160.

Lipman, Z. (2011). *Trade in hazardous waste: Environmental justice versus economic growth.* Available at: https://www.oceanfdn.org/sites/default/files/Trade%20in%20Hazardous%20Waste.pdf. Last accessed 6 Oct 2021.

Looking North: Sharpening America's Arctic focus. (2019, May 6). *Speech. Michael R. Pompeo, secretary of state.* Rovaniemi, Finland. URL: https://www.state.gov/looking-north-sharpening-americas-arctic-focus/. Accessed 14 Nov 2020.

Lozovik, P. A. (2006). The stability of water bodies to acidification depending on their specific catchment on the example of soils and lakes of the Shui River basin (Onega). *Water Resources, 2*, 188–194.

Lukin, Y. F. (2011). Solovetsky Forum-2011. *Arctic and North, 3*, 7–14.

Dongwei, L., Yua, Q., Danni, Z., Xie, J., Xiaodong, G., Gaoyue, S., Bin, Z., Wang, S., Thorjørn, L., & Lei, D. (2022). Critical loads of headwater streams in China using SSWC model modified by comprehensive F-factor. *Science of the Total Environment, 802*, 149780. https://doi.org/10.1016/j.scitotenv.2021.149780

Ma, Y. (2020). *Marine oil and gas exploration in China* (p. 581). Springer-Verlag.

Markelov, V. A., Andreev, O. P., Kobylkin, D. N., Arabsky, A. K., Arno, O. B., Tsybulsky, P. G., Bashkin, V. N., Kazak, A. S., & Galiulin, R. V. (2013). *Gas industry sustainable development* (p. 211). Nedra.

Martinez-Alier, J. (2001). Mining conflicts, environmental justice, and valuation. *Journal of Hazardous Materials, 86*, 153–170.

Matishov, G. G., & Dezhnyuk, S. L. (2007). Studies of the Arctic seas along the Northern Sea route. In G. G. Matishov (Ed.), *Biology and oceanography of the Northern Sea route: The Barents and Kara seas* (pp. 5–28). Nauka.

Matten, D., & Crane, A. (2005). Corporate citizenship: Towards an extended theoretical conceptualization. *The Academy of Management Review, 30*(1), 166–179.

Melas, M. (2016). The Arctic as a geopolitical bond among the European Union, Norway & Russia. In L. Heininen, H. Exner-Pirot, & J. Plouffe (Eds.). The Arctic yearbook: 20 years of regional cooperation and policy-shaping). Northern Research Forum/University of the Arctic (pp. 326–337).

Meshcherin, I. V., Kazak, A. S., Bashkin, V. N., & etc. (2011). *Globalization of the natural gas market* (p. 348). Gazprom-VNIIGAZ.

Mingulov, R. M., Raykevich, S. I., & Fokina, L. M. (2005). New technologies for reduction of nitrogen oxides emission in atmosphere. *Gas Industry, 5*, 74–76.

Mitko, A. V. (2016). Features of the Arctic calls Russian geopolitics. *Digest of World Politics. Annual Review, 8*, 173–180.

Mitko, A. V. *Features of the arctic challenges of Russian geopolitics*. https://studref.com/420639/politologiya/osobennosti_arkticheskih_vyzovov_rossiyskoy_geopolitiki. Accessed 14 Nov 2020.

Moiseenko, T. I. (2003). *Sustainability of fresh water ecosystems to acid deposition in the European Russia* (p. 243). Nauka.

Moiseenko, T. I. (2005). Acidification of waters and conjugate behavior of elements of the chemical composition of waters. *Geochemistry, 10*, 1120–1127.

Moiseenko, T. I., Dinu, M. I., Bazova, M. M., & de Wit, H. A. (2015). Long-term changes in the water chemistry of arctic lakes as response to reduction of air pollution: Case study of the Kola, Russia. *Water, Air, and Soil Pollution, 226*, 97–104.

Moiseenko, T. I., Gashkina, N. A., & Dinu, M. I. (2016). *Water acidification: Vulnerability and critical loads* (p. 393). Lenand.

Moiseenko, T. I., Gashkina, M. M., & Dinu, M. I. (2017). Enrichment of surface water by elements: effect of air pollution, acidification and euthrophication. *Environmental Process, 1*, 3–10.

Morgunov, B. A. (2011). *Diagnostic analysis of the ecological state of the Russian Arctic* (p. 172). Nauchnyy Mir Publ.

National Security Presidential Directive (NSPD-66) and Homeland Security Presidential Directive (HSPD-25). (2009, Jan 12). URL: https://polarconnection.org/national-security-presidential-directive-66homeland-security-presidential-directive-25-january-2009/. Accessed 10 Oct 2020.

Nilsson, J., & Grennfelt, P. (1988). *Critical loads for sulfur and nitrogen*. NORD Status Report: 15. Nordic Council of Ministers, Copenhagen, Denmark.

Nixon, S. W., & Pilson, M. E. Q. (1983). Nitrogen in estuarine and coastal marine ecosystems. In E. J. Carpenter & D. G. Capone (Eds.), *Nitrogen in the marine environment* (pp. 565–290). Academic Press.

Nozik, R. (1974). *Anarchy, state and utopia* (p. 560). Basic Books.

Nuryshev, G. N. (2012). Arctic challenges of Russian geopolitics. *Geopolitika i bezopasnost; [Geopolitics and Security]., 2*(18), 83–89.

Ocampo José Antonio. (2013). *Super-cycles of commodity prices since the mid-nineteenth century*. Presentation at the International Monetary Fund. https://www.imf.org/external/np/seminars/eng/2012/commodity/pdf/Ocampo.pdf. Accessed 04 Feb 2021.

Onishchenko, G. G., Novikov, S. M., Rakhmanin, Y. A., Avaliani, S. L., & Bushtueva, K. A. (2002). *Fundamentals of risk assessment for public health under the influence of chemicals that pollute the environment* (p. 408). Research Institute Echigos.

Pickford, A., & Collins, J. F. (2016). *Toward a new polar partnership: A framework for Canadian-Australian cooperation* (p. 23). Macdonald-Laurier Institute.

Pinto, R., Patricio, J., Baeta, A., Fath, B. D., et al. (2009). Review and evaluation of estuarine biotic indices to assess benthic condition. *Ecological Indicators, 9*, 1–25.

Priputina, I. V., & Bashkin, V. N. (2017). Assessing critical loads for polar marine ecosystem. In V. Bashkin (Ed.), *Ecological and biogeochemical cycling in impacted polar ecosystems* (pp. 237–248). NOVA Science Publishers.

Rawls, J. (1972). *A theory of justice* (p. 560). Harward University Press.

Risk Analysis. (1988). *A guide to principles and methods for analyzing health and environmental risk*. NTIS: ND.

Risk and Uncertainty in Environmental Impact Assessment. Environmental Impact Assessment for Developing Countries in Asia. Volume 1: Overview. (1997). B. N. Lohani, J.W. Evans, R.R. Everitt, et al. (Eds.). http://www.asiandevbank.org/Documents/Books/Environment_Impact/chap5.pdf

Robinson, P. B. (2002). All for some: Water inequity in Zambia and Zimbabwe. *Physics and Chemistry of the Earth, 27*(11–22), 851–857.

Romankevich, E. A., & Vetrov, A. A. (2001). *Carbon cycle in the Arctic seas of Russia* (p. 301). Nauka.

Rosén, K., Gundersonv, P., & Tugnhammar, L. (1992). Nitrogen enrichment in northern forest ecosystems. *Ambio, 21*, 364–368.

Ross, E. (2017, July 14). How China's Arctic Empire Will Upset the Global Balance of Power. *Newsweek.* URL: http://www.newsweek.com/chinas-arctic-goals-620794

RSPP. (2015). *National register and library of non-financial reporting*. URL: http://www.rspp.ru/simplepage/157. Accessed 22 Oct 2021.

Rusakova, V. V., Kazak, A. S., Bashkin, V. N., Bukhgalter, E. B., Galiulin, R. V., & Akopova, G. S. (2009). *Environmental risk management in the gas industry* (p. 200). Gazprom-VNIIGAZ.

Samsonov, R. O., Kazak, A. S., Bashkin, V. N., & Lesnykh, V. V. (2007a). *System analysis of geo-ecological risks* (p. 272). Scientific World.

Samsonov, R. O., Kazak, A. S., & Bashkin, V. N. (2007b). *Master plan methodology for gas industry development* (p. 304). Scientific World.

Sardanashvili, S. A. (2005). *Calculation methods and algorithms (pipeline gas transportastion)* (p. 223). Oil and Gas.

Semenov, M. Y. (2002). *Acid precipitation on the territory of Siberia: Calculation and mapping of permissible loads* (p. 144). Nauka.

Sen, A. (2000). *Freedom, social choice and responsibility: Arrow lectures and other essays* (p. 400). Clarendon Press.

Sergunin, A. A. (2021). Strategies of corporate social and environmental responsibility in the Russian Arctic: Theoretical aspects. *Greater Eurasia: Development, security, cooperation. Yearbook. Issue 4, Part 1*, 732–735.

Sharkey, A. J., & Bromley, P. (2015). Can ratings have indirect effects? Evidence from the organizational response to peers' environmental ratings. *American Sociological Review, 80*, 63–91.

Sheppard, B. H., Lewicki, R. J., & Minton, J. W. (1992). *Organizational justice: The search for fairness in the workplace* (p. 227). Lexington Books.

Shvarts, E. A., Kniznikov, A. Y., Pakhalov, A. M., & Sheresheva, M. Y. (2015). Assessment of environmental responsibility of oil and gas companies in Russia: The rating method. *MSU Vestnik, Series 6, Economics, 5*, 46–67.

Shvarts, E. A., Pakhalov, A. M., & Kniznikov, A. Y. (2016). Assessment of environmental responsibility of oil and gas companies in Russia: The rating method. *Journal of Cleaner Production, 127*, 143–151.

Sigareva, L. E. (2013). *Chlorophyll in bottom sediments of the Volga water bodies* (p. 217). KMK.

Smirnov, A. I. (2016). *The Arctic: Network diplomacy 2.0 in the global security discourse* (p. 157). NArFU Publ.

Sobol, I. M. (1985). *The Monte Carlo method* (p. 64). Nauka.

Statement on Canada's Arctic Foreign Policy. Exercising Sovereignty and Promoting Canada's Northern Strategy Abroad. (2010). (p. 27).

Stammler, F., & Peskov, V. (2008). Building a «culture of dialogue» among stakeholders in North-west Russian oil extraction. *Europe-Asia Studies, 60*(5), 831–849.

Stymne, S., & Jackson, T. (2000). Intra-generational equity and sustainable welfare: A time series analysis for the UK and Sweden. *Ecological Economics, 33*, 219–236.

The Red Book of the Yamalo-Nenets Autonomous Okrug: animals, plants, fungi. (1997). L. N. Dobrinsky (Ed.). Ural University Press (p. 240).

Redfield, A. C. (1958). The biological control of chemical factors in the environment. *American Science, 46*, 823–847.

Trubitsina, O. P. (2015). Ecological monitoring of acid deposition in the Arctic region. *Open Ecology Journal, 8*, 21–31.

Trubitsina, O. P. (2016a). Biota monitoring in the impacted zones of oil and gas industry in the Arctic region. In V. N. Bashkin (Ed.) *Biogeochemical technologies for managing environmental pollution in polar ecosystems* (pp. 87–94).

Trubitsina, O. P. (2016b). Acid-alkaline features of precipitation in the coastal zone of the Arkhangelsk region: a retrospective analysis. Bulletin of the Northern (Arctic) federal university. *Series: Natural Sciences, 4*, 17–25. https://doi.org/10.17238/issn2227-6572.2016.4.17

Trubitsina, O. P., & Bashkin, V. N. (2016). Analysis of geoecological risks and ratings as a factor of increasing the investment attractiveness of enterprises. *Issues of Risk Analysis, 13*(3), 76–83.

Trubitsina, O. P., & Bashkin, V. N. (2017a). Geoecology and geopolitic in the arctic region: Ecological and political risks and challenges. In V. Bashkin (Ed.), *Ecological and biogeochemical cycling in impacted polar ecosystems* (pp. 217–236). NOVA Science Publishers.

Trubitsina, O. P., & Bashkin, V. N. (2017b). Environmental ratings as a factor of improving investment attractiveness of Russian oil and gas companies, operating in the Arctic. In V. Bashkin (Ed.), *Ecological and biogeochemical cycling in impacted polar ecosystems* (pp. 275–292). NOVA Science Publishers.

Trubitsina, O. P., & Bashkin, V. N. (2018). Challenges of the activity of oil and gas industry facilities in the Arctic: Geoecological and geopolitical risks. *Issues of Risk Analysis, 15*(3), 66–76.

Trubitsina, O. P., & Bashkin, V. N. (2019a). Geoenvironmental risks on the background geopolitical challenges for the oil and gas industry in the Arctic. *Neftegaz.ru, 4*, 78–82.

Trubitsina, O. P., & Bashkin, V. N. (2019b). Environmental rating as an indicator of geoecological risk management of Russian oil and gas companies in the Arctic. *Issues of Risk Analysis, 16*(2), 58–69. https://doi.org/10.32686/1812-5220-2019-16-58-69

Trubitsina, O. P., & Bashkin, V. N. (2019c). Geoecological risk against the background of geopolitical challenges of the oil and gas industry in the Arctic. *Issues of Risk Analysis, 16*(4), 12–23. https://doi.org/10.32686/1812-5220-2019-16-4-12-23

Trubitsina, O. P., & Bashkin, V. N. (2019d). Taking into account geopolitical challenges in the analysis of environmental risks of oil and gas development of the Russian Arctic. *Issues of Risk Analysis, 16*(6), 50–59. https://doi.org/10.32686/1812-5220-2019-16-6-50-59

Trubitsina, O. P., & Bashkin, V. N. (2020). Dynamics of environmental responsibility of oil and gas industry facilities in the Russian Arctic: A rating approach for 2014–2019. *Issues of Risk Analysis, 17*(6), 10–21. https://doi.org/10.32686/1812-5220-2020-17-6-10-21

UBA. (1996). *Manual on methodologies and criteria for mapping critical levels/loads and geographical areas where they are exceeded.* UNECE convention on long-range transboundary air pollution, Federal Environmental Agency, Berlin.

UBA. (2004). *Mapping Manual for modelling and mapping of critical loads and their input data'2004.* http://www.icpmapping.org

United Nations, Department of Economic and Social Affairs, Population Division. (2019). *World population prospects 2019: Highlights.* URL: https://population.un.org/wpp/Publications/Files/WPP2019_Highlights.pdf

US National Strategy for the Arctic: A Path to Cooperation. International Expert Council on Cooperation in the Arctic. (2013). URL: http://www.iecca.ru/zakonodatelstvo/voprosy-prava/item/146-natsionalnaya-strategiya-ssha-dlya-arktiki-put-k-sotrudnichestvu. Accessed 10 Oct 2021.

Vadstein, O., Harkjerr, B. O., & Jensen, A. (1989). Cycling of organic carbon in the photic zone of an eutrophic lake with special reference to heterotrophic bacteria. *Limnology and Oceanography, 34*, 840–855.

VandeVeer, D., & Regan, T. (1987). *Health care ethics: An introduction* (p. 464). Temple University Press.

Vandysh, O. I. (2002). The effect of acidification on zooplankton communities of small lakes of the mountain tundra. *Water Resources, 5*, 602–609.

Varian, H. R. (1974). Equity, envy and efficiency. *Journal of Economic Theory, 9*(1), 63–91.

Vet, R., Artz, R. S., Carou, S., Shawa, M., Ro, C., Aas, W., et al. (2014). A global assessment of precipitation chemistry and deposition of sulfur, nitrogen, sea salts, base cations, organic acids, acidity and pH and phosphorus. *Atmospheric Environment, 93*, 3–100.

Vitousek, P. M., & Howard, R. W. (1991). Nitrogen limitation on land and in the sea. *Biogeochemistry, 13*, 87–115.

Wernstedt, K. (2002). *Environmental Management in the Russian Federation: A next generation enigma* (p. 37). Resources for the Future.

Willemse, C. A., & Van Gelder, P. H. A. J. M. (2011, Jan). *Oil pollution in the Arctic: Risks and mitigating measures with reference to the deepwater horizon accident.* Conference: ASME 2011 30th International Conference on Ocean, Offshore and Arctic Engineering. https://doi.org/10.1115/OMAE2011-49157

Williams, R. H. (1999). Environmental justice in America and its politics of scale. *Political Geography, 18*, 49–73.

World Energy Outlook 2021. (2021). IEA. 386p. URL: https://iea.blob.core.windows.net/assets/88dec0c7-3a11-4d3b-99dc-8323ebfb388b/WorldEnergyOutlook2021.pdf. Accessed 21 Oct 2021.

Wulff, F. (1990). *Large-scale environmental effects and ecological processes in the Baltic Sea.* Swedish EPA, Report N3849.

Yashalova, N. N., & Ruban, D. Y. (2014). The special significance of the environmental factor for the sustainable development of the national economy: A conceptual analysis. *Economy Development Strategy, 14*(251), 20–30.

Zenkevich, L. A. (1963). *Biology of the seas of the USSR* (p. 740). Izd-vo AS USSR.

Zhiznin, S. Z., & Timokhov, V. M. (2019). Economic and geopolitical aspects of the Nord Stream-2. *Baltic Region, 11*(3), 25–42.

Index

V. N. Bashkin, O. P. Trubitsina, *Geoecological and Geopolitical Risks for the Oil and Gas Industry in the Arctic*, Environmental Pollution 29,
https://doi.org/10.1007/978-3-030-95910-4